SOIL METAGENOMICS

Microbes rule the world. Unveiling the microbiome is a challenge. It was evident that only < 2% of microbes in the globe is culturable and the remaining is hidden. Hence the current knowledge on existing microbiome is scarce. In the era of modern omics, metagenomics has spawned up as a powerful tool to unlock this blackbox. Today metagenomics in microbiology has travelled long in environmental and clinical microbiology across several barriers. Functional metagenomics offers sustainable solutions to planet, people and profit to bio-economy through high resolution 4G sequencing techniques. Data integration is a crucial step in metagenomics data analysis. Viral metagenomics is a least explored one. The emphasis on cutting edge research seems integration of high throughput sequencing and bioinformatics. Concisely considering the significance of soil metagenome, **Soil Metagenomics** provides a concrete base for understanding the principles and methods involved in soil metagenomics. This book focuses on the recent advents and technological breakthroughs in metagenomic approaches coupled with their applications in agriculture. The intended audience include soil and environmental microbiologists, molecular biologists and policy makers. The book

- Expertly describes the latest fourth generation metagenomic technologies from sample collection to data analysis, metatranscriptomic, metaproteomic and metabolomics studies
- Includes miniquiz and a glossary session for easy understanding of the readers
- Presented topics include techniques in soil metagenomics, gene discovery, taxanomic and functional diversity of soil microbial communities, mobile metagenomics, viral metagenomics, bioinformatics and data analysis, fourth generation sequencing, application of metagenomics in agriculture and the future of metagenomics.

Dr. T.C.K. SUGITHA obtained her Ph.D. and PDF from Tamil Nadu Agricultural University, Tamil Nadu, India. She served as ARS Scientist in Indian Agricultural Research Institute, New Delhi and Central Potato Research Institute, Shimla.

Dr. ASISH K BINODH is Assistant Professor of Plant Breeding and Genetics in Agricultural College and Research Institute, Killikulam, sister campus of Tamil Nadu Agricultural University, Coimbatore, India.

Prof. Dr. K. RAMASAMY received his Bachelors and Masters from Annamalai University. He later pursued his Master of Science (M.S) in Fermentation Technology at Catholic University of Leuven, Belgium and PhD research in the field of Industrial Microbiology.

Prof. SIVAKUMAR UTHANDI is Professor of Agricultural Microbiology in Tamil Nadu Agricultural University, Coimbatore, India.

SOIL METAGENOMICS

T.C.K. Sugitha
Asish K. Binodh
K. Ramasamy
U. Sivakumar

NARENDRA PUBLISHING HOUSE
DELHI (INDIA)

First published 2021
by CRC Press
2 Park Square, Milton Park, Abingdon, Oxon, OX14 4RN

and by CRC Press
6000 Broken Sound Parkway NW, Suite 300, Boca Raton, FL 33487-2742

CRC Press is an imprint of Informa UK Limited

British Library Cataloguing-in-Publication Data
A catalogue record for this book is available from the British Library

Library of Congress Cataloging-in-Publication Data
A catalog record has been requested

ISBN: 978-0-367-69396-1 (hbk)
ISBN: 978-1-003-14167-9 (ebk)

CONTENTS

PREFACE

Microorganisms play an integral and unique role in ecosystem function and sustainability. Most microbes live in highly organized and interactive communities that are versatile, complex, and difficult to analyze from many perspectives. Understanding the structure and composition of microbial communities and their responses and adaptations to environmental perturbations such as toxic contaminants, climate change, and agricultural and industrial practices is critical in maintaining or restoring desirable ecosystem functions. However, because less than 1% of microorganisms have been cultivated, characterization and detection of microbial populations in natural environments present a great challenge to microbial ecologists.

Current methods for analyzing microbial communities, especially their key functions, are too cumbersome. Rapid, simple, reliable, quantitative, and cost-effective tools that can be operated in real-time and in heterogeneous field-scale environments are needed. The microbial world encompasses millions of genes from thousands of species, with hundreds of thousands of proteins and multi molecular machines operating in a web of hundreds of interacting processes in response to numerous physical and chemical environmental variables. Gene control is complex, with groups or "cassettes" of genes (operons) directing coordinated transcription and translation of genes into interacting proteins. The ability to extract and characterize genomic DNA fragments from mixed microbial assemblages in environmental samples is providing novel insights into the ecology, evolution, and metabolism of uncultured microorganisms in nature.

Metagenomics is a rapidly growing area of the genome sciences that seeks to define the features of intact microbial communities in their native habitats. It allows us to see, with a new and powerful set of lenses, the vast microbial and metabolic diversity that exists in our biosphere. The scale of this diversity is mind-boggling. The estimated total number of microbial cells is 10^{30} and may harbor as many as 10^{31} phages. Their genomes encode a wide variety of enzymes that influence major metabolic fluxes within the biosphere, from phototrophy in the sea and , to phosphate removal in industrial sludge, to nutrient harvest from the diet in animal guts. Metagenomics provides us with an unprecedented opportunity to assess the metabolic features of microbial communities without the need to culture their component members. The results promise to impact a wide range of "bio"-related disciplines, including biomedicine, bioenergy, bioremediation, biodefense and agriculture.

Metagenomics enabled the microbial ecologists to solve the unresolved puzzle of unculturables and the soil biologists to open the soil black box. Numerous practical implications add excitement to this field and the volume of work performed is increasing integrated with other 'omics'. The advent of next generation sequencing technologies created a paradigm shift in metagenomics in exploring the diversity and functions of these unsung heroes of soil. Though many manuscripts are available for updating our knowledge on Metagenomics in a general perspective, the implications of this dynamic field in relevance to agriculture is lacking. Understanding the importance of soil metagenomics in agriculture, this masterpiece is the brainchild of our working group with greater emphasis to recent trends in soil metagenomics.

We the authors are much grateful to present this manuscript '*Glimpses of soil metagenomics*' to the student community, teachers and researchers and hope that it will provide an overview of this interesting field and stimulate the researches to add colors with new discoveries. The glossary is intended to aid the reader in finding new terms that are used in metagenomics. Throughout the text book, we discuss not only the state of our understanding but also the methodology employed.

We whole heartedly thank the 'Almighty' who is the author and creator of everything inclusive of these magic microbes. We wish to acknowledge our colleagues who have assisted in the development of this manuscript. We are again grateful to our families for their support during the writing of this book.

Authors

CHAPTER - 1

INTRODUCTION TO METAGENOMICS

Soil is a living organ of the earth. Until recent times large and small civilizations throughout the world recognized this truth, expressing it through a deep spiritual relationship with soil. The Old Testament in Bible describes the soil as a source of healing. ***"The Lord hath created medicines out of the earth, and he that is wise will not abhor them" (Ecclesiasticus 38:4).*** Even in New Testament of Bible narrates an incident where Jesus Christ healed a blind man by putting a paste made from soil on the man's eyes (John 9). The ancient Greeks worshipped soil through the goddesses Gaia and Demeter, the Germans through goddesses Ertha and the Native Americans through Mother Earth.

Advances in science, medicine and agriculture during the 20th century provided numerous reasons to continue to venerate the soil well beyond the basic necessity of food production. Discovery during this century about soil biology led to the development of life saving drugs including antibiotics, antitumour agents and immune-suppressants. And yet, just as these advances grew more impressive, urbanization and other global economic influences tended to disassociate many people from an intimate association with the environment and gradually we lost appreciation of our reliance on the soil. But that trend may be altered by the renaissance in soil biology that presents the soil as one of the last great frontiers available for discovery and bioprospecting. This renaissance is driven by the tools of molecular biology, which enable exploration of the life and chemistry of soil in new ways. One emerging tool that draws together traditional soil biology and modern molecular biology is metagenomics which entails collective analysis of genomes of an assemblage of microorganisms.

1.1. SOIL AS A MICROBIAL HABITAT

Soil comprises mineral particles of different sizes, shapes and chemical characteristics, together with the soil BIOTA and organic compounds in various stages of decomposition. The biological processes in soil are executed in a complex physical and chemical environment. The physical structure of soil is varied and dynamic soil particles vary in size from 1 to 1000 micrometers and the moisture content and chemistry of soil affect particle size. The particle size inturn affects water flow, gas exchange and temperature gradient in soil. The chemical composition of soil is derived from a combination of its geologic and biologic origins.

The formation of clay–organic matter complexes and the stabilization of clay, sand and silt particles through the formation of aggregates are the dominant structural characteristics of the soil matrix. Soil-matrix-component aggregates range from approximately 2 mm or more (macro aggregates) to fractions of a micrometer for bacteria and colloidal particles.

Prokaryotes are the most abundant organisms in soil and can form the largest component of the soil biomass. Soil microorganisms often strongly adhere or adsorb onto soil particles such as sand grains or clay–organic matter complexes. Microhabitats for soil micro organisms include the surfaces of the soil aggregates, and the complex pore spaces between and inside the aggregates. Some pore spaces are inaccessible for microorganisms owing to size restrictions.

1.1.1. Soil moisture and Microbial biomass

The metabolism and the survival of soil microorganisms are strongly influenced by the availability of water and nutrients. In contrast to aquatic habitats, surfaces of soil environments undergo dramatic cyclic changes in water content, ranging from water saturation to extreme aridity. A fraction of the microbial community dies during each drying-and-wetting cycle. As a consequence, the composition of soil microbial communities fluctuates.

1.1.2. Microbial biomass and biogeochemical cycles

Microorganisms active in the soil are largely responsible for biogeochemical cycles that support life on earth. Soil is an important reservoir for organic carbon, and prokaryotes are an essential component of the soil decomposition system. Despite the high concentration of organic matter in most soil types, only low concentrations of organic carbon are readily available to microorganisms. Reasons for this include the transformation of most of the organic matter that is derived from plants, animals and microorganisms into humus by a combination of microbiological and abiotic processes, and the uneven distribution of microorganisms and organic compounds in the soil matrix. Humic substances are stable and recalcitrant to microbial decomposition processes the half-life of these stable organic matter complexes with respect to biological degradation is approximately 2,000 years. Since many of these changes are driven by microorganisms the structure and function of the microbial communities get affected.

Apart from Carbon cycle in soil, microorganisms also mediate the Nitrogen cycle, Sulphur cycle, oxidative and reductive transformation of metals such as iron, silica and mercury. Soil microorganisms also influence the health of plants and animals by providing vitamins and other nutrients, thereby influencing the developmental process. Microorganisms exist within a carefully balanced natural ecosystem of checks and balances and when disrupted through human agency can produce harmful environmental effects. Hence the microbial diversity and the corresponding gene pool, must be large.

1.2. HISTORY OF SOIL METAGENOMICS

Ever since Robert Koch established his pioneering postulates on the microbial nature of disease, the field of microbiology has centered on the process of cultivating individual microbial species. Growing an organism in pure culture has been the critical first step towards understanding the properties of a given microbe. This culture dependent, reductionist approach has produced many impressive successes of microbiology in the 20th century. However, the limitations of culture dependent studies were first recognized by what has become known as the 'great plate count anomaly'. The advent of new molecular and microscopic methods has fostered a growing realization, however, that there is much to be known about soil microorganisms that cannot be elucidated through traditional approaches. At the heart of this revolution was the convincing demonstration that the uncultured microbial world far outsized the cultured world and that this unseen world could be studied.

In 1931, Waksman optimistically believed that "a large body of information has accumulated that enables us to construct a clear picture of the microscopic population of the soil", and in 1923 *Bergey's Manual* stated categorically that no organism could be classified without being cultured. Pace and colleagues highlighted the need for nontraditional techniques to understand the microbial world: "The simple morphology of most microbes provides few clues for their identification;

physiological traits are often ambiguous. The microbial ecologist is particularly impeded by these constraints, since so many organisms resist cultivation, which is an essential prelude to characterization in the laboratory". In the ensuing years, microbiologists dedicated intense effort to describe the phylogenetic diversity of exotic and ordinary environments ocean surfaces, deep sea vents, hot springs,soil, animal rumen and gut, human oral cavity and intestine. Many new lineages were classified based on their molecular signatures alone. The next challenge was to elucidate the functions of these new phylotypes and determine whether they represented new species, genera, or phyla of prokaryotic life. This challenge spawned various techniques, including metagenomics, the genomic analysis of assemblages of organisms. In a few years, the study of uncultured microorganisms has expanded beyond asking "Who is there?" to include the difficult question "What are they doing?." The recent advent of pyrosequencing and next generation sequencing technologies unlocked the magics and menaces of microbes present in a given environment. Some of the challenges that opened the concept of 'uncultured world' are discussed below.

1.2.1. *Sulfolobus* from Yellowstone hot springs

From the 1880s forward, the microbiological world was divided into the cultured and the uncultured. It is very obvious that culturing did not capture the full spectrum of microbial diversity. One of the indicators that cultured microorganisms did not represent much of the microbial world was the oft-observed "great plate count anomaly" the discrepancy between the sizes of populations estimated by dilution plating and by microscopy. Brock and colleagues encountered microorganisms in Yellowstone hot springs that could not be cultured and others whose behavior in culture did not reflect their activities *insitu*. Due to the Yellowstone Plateau's high elevation the average boiling temperature at Yellowstone's geyser basins is 199 °F (93 °C). Water erupting from Yellowstone's geysers is superheated above that boiling point to an average of 204 °F (95.5 °C) as it leaves out the vent. Many of the organisms could not be cultured on agar medium because their temperature requirements exceed the melting point of the agar. Brock's central technique involved the immersion of microscope slides in the spring for 1 to 7 days, followed by microscopic examination and often staining with fluorescent antibodies raised against cultured members of the taxonomic groups suspected to inhabit the environment. This approach estimated *in situ* population sizes and growth rates, which indicated, for example, that certain strains of *Sulfolobus* grew in the hot springs at temperatures well below the optima in pure culture. Further Brock's group inferred that the genus

Synechococcus, typically pink in culture, was a major contributor to photosynthesis in hot springs.

1.2.2. *Vibrio cholera* – the challenging food pathogen

Further evidence that drew attention to the uncultured world accumulated during the 1970s and 1980s. A study of oligotrophs indicated that incubation times longer than 25 days enhanced the recovery of certain organisms in culture. The food industries generated intense interest in "injured bacteria" in food live organisms that cannot be cultured following stressful treatments such as heat, chilling, or desiccation but represent a significant risk to human health The concept of organisms that were viable but not culturable emerged from the work of Colwell and colleagues, who showed that strains of *Vibrio cholerae* were indeed alive and virulent when isolated from aquatic environments but did not grow in culture until after passage through a mouse or human intestine.

1.2.3. The dreadly *Helicobacter Pyroli*

The confluence of these and many other scientific and technical advances steadily drew attention to the unculturable microbial world, but two discoveries figured significantly in the sharpened focus. i)The diversity of soil bacteria, which demonstrated with DNA-DNA reassociation techniques confirmed that the complexity of the bacterial DNA in the soil was at least 100-fold greater than could be accounted for by culturing. ii)The second discovery was the demonstration that *Helicobacter pylori* causes gastric ulcers and cancer. Although spiral bacteria had been observed in the gastric mucosa of dogs in 1893 and in humans in 1906, the correlations between the appearance of the bacteria and peptic ulcers were noted in 1938 untill *H. pylori* was cultured. These discoveries provided compelling evidence that drew microbiologists to wrestle with the daunting challenge of devising strategies to access these organisms.

1.2.4. A shift in the Paradigm

In 1985, an experimental advance radically changed the way we visualize the microbial world. Building upon the pioneering work of Carl Woese, which showed that rRNA genes provide evolutionary chronometers, Pace and colleagues created a new branch of microbial ecology. They used direct analysis of 5S and 16S rRNA gene sequences in the environment to describe the diversity of microorganisms in an environmental sample without culturing.

The next technical breakthrough arrived with the development of PCR technology and the design of primers that can be used to amplify almost the entire gene. This accelerated the discovery of diverse taxa as habitats across the earth. The application of PCR technology provided a view of microbial diversity that was not distorted by the culturing bias and revealed that the uncultured majority is highly diverse and contains members that diverge deeply from the readily culturable minority. Today, 52 phyla have been delineated, and most are dominated by uncultured organisms (Fig. 1.1).

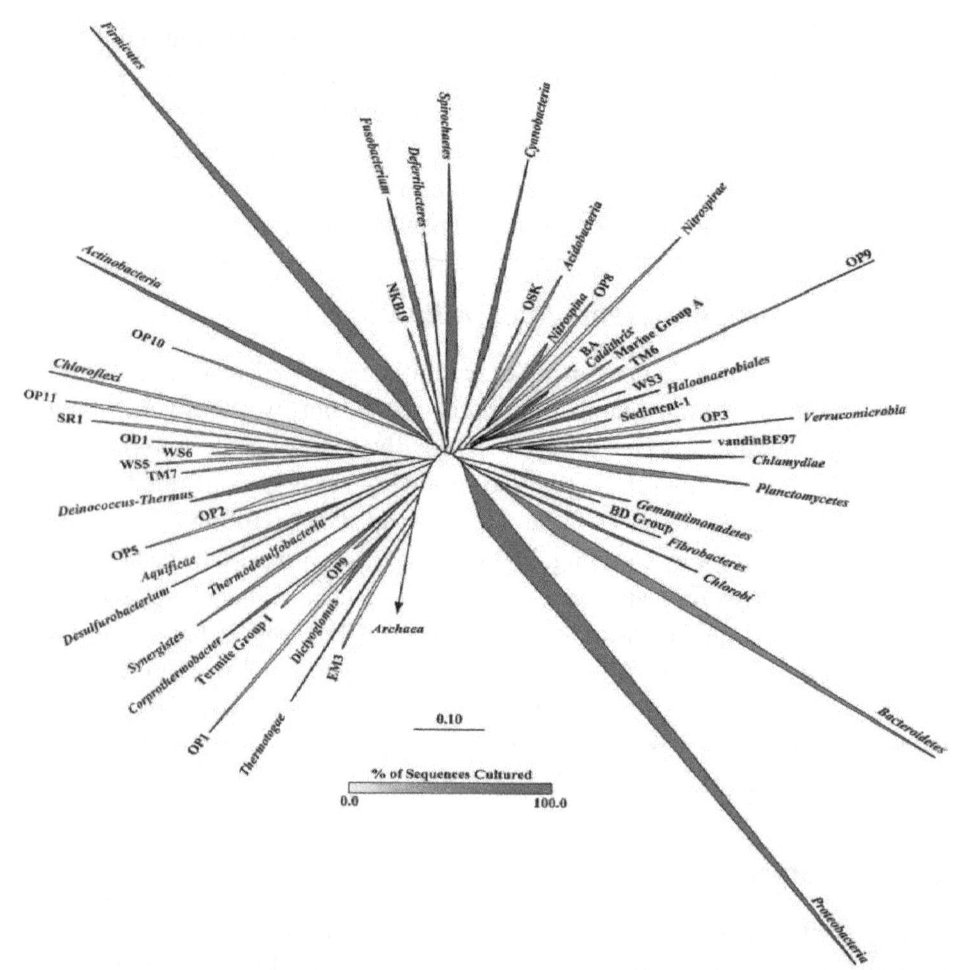

Fig. 1.1. Phylogenetic tree of *Bacteria* showing established phyla (italicized Latin names) and candidate phyla .The vertex angle of each wedge indicates the relative abundance of sequences in each phylum; the length of each side of the wedge indicates the range of branching depth found in that phylum; the redness of each wedge corresponds to the proportion of sequences in that phylum obtained from cultured representatives. Candidate phyla do not contain any cultured members.

The application of phylogenetic stains (Fig. 1.2) nucleic acid probes with fluorescent labels facilitate visualization of single cells *insitu* where as traditional microscopy provides little phylogenetic information. Fluorescent antibody studies require prior knowledge and culturing of an organism or one closely related to it to raise antibodies. Phylogenetic stains require only an rRNA sequence, which can be derived from an environmental sample without culturing.

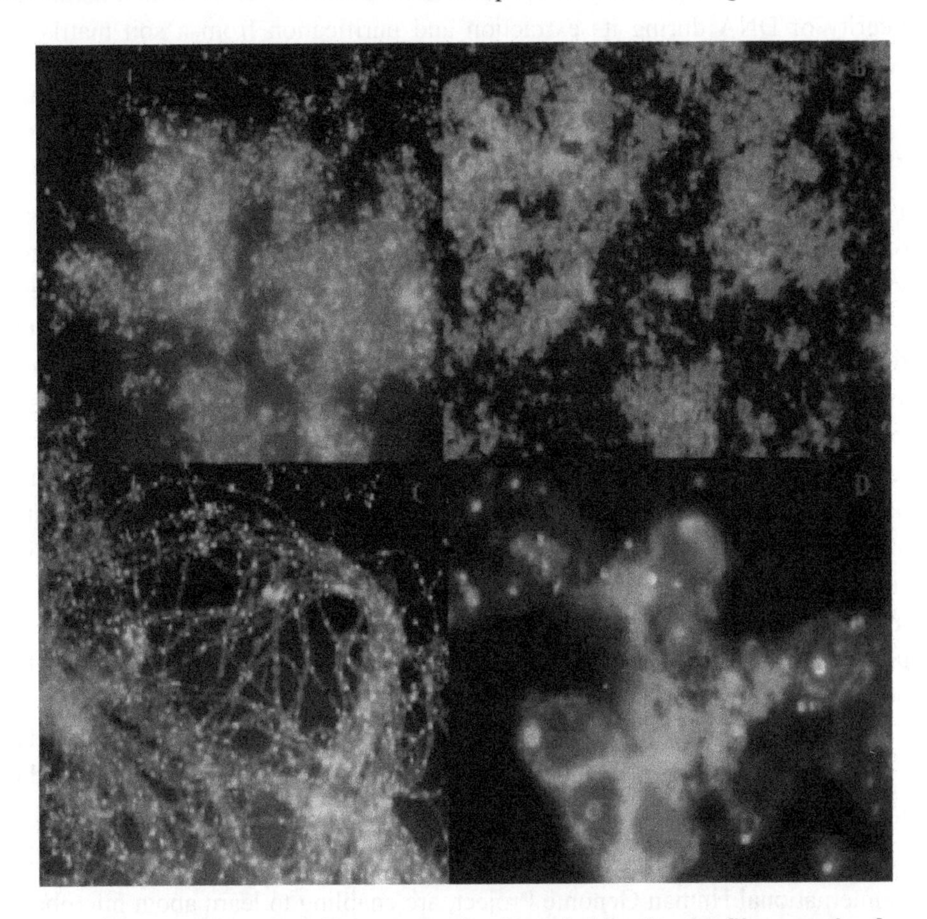

Fig. 1.2. Phylogenetic stains. Fluorescent in situ hybridization biofilm samples from Iron Mountain Mine, Calif. Nucleic acid probes were labeled with indodicarbocyanine, and DNA was stained nonspecifically with 4_,6_-diamidino-2-phenylindole. The nucleic acid probes are specific for (top left) *Sulfobacillus* spp., (top right) *Archaea*, (bottom left) *Archaea* on fungal filaments, and (bottom right) *Eukarya*.(Courtesy:Baker *etal*., 2003)

Pace and his colleagues during the year 1991, first proposed the idea of cloning DNA directly from environmental samples and in 1991, the first such cloning in a phage vector was reported. The next advance was the construction of a metagenomic library with DNA derived from a mixture of organisms enriched on dried grasses. Clones expressing cellulolytic activity were found in these libraries, which were referred to as zoolibraries. The work of DeLong's group reported libraries constructed from prokaryotes in seawater and identified a 40-kb clone that contained a16S rRNA gene indicating that the clone was derived from an archaeon that had never been cultured. Construction of libraries with DNA extracted from soil lagged due to difficulties associated with maintaining the integrity of DNA during its extraction and purification from a soil matrix but eventually produced analyses analogous to those from seawater.

1.2.5. rRNA Analysis and Culturing

In addition to providing a universal culture-independent means to assess diversity, 16S rRNA sequences also provided an aid to culturing efforts. Bacteria remains recalcitrant to culturing for diverse reasons—lack of necessary symbionts, nutrients, or surfaces, excess inhibitory compounds, incorrect combinations of temperature, pressure, or atmospheric gas composition, accumulation of toxic waste products from their own metabolism, and intrinsically slow growth rate or rapid dispersion from colonies. Testing myriad conditions require focus on the critical variables, are challenging and laborious, and can only succeed if there is a sufficiently quantitative assay available to determine whether the organism of interest is enriched under a specific set of conditions. Nucleic acid probes labeled with fluorescent tags facilitate quantitative assessment of enrichment and growth. As a result, culturing efforts have intensified recently, and successes have included pure cultures of members of the genus *Pelagibacter* which represents more than one-third of the prokaryotic cells in the surface of the ocean but was known only by its 16S rRNA signature until 2002. The phylum *Acidobacteria* are abundant in soil, typically representing 20 to 30% of the 16S rRNA sequences amplified by PCR from soil DNA.

Technological advances, particularly in genetic research conducted as part of the international Human Genome Project, are enabling to learn about microbes at their most fundamental level. The challenges are accepted by scientific thought and offer beneficial applications in areas important to DOE's (U.S. Department of Energy) Biological and Environmental Research (BER) program *viz.*, bioremediation, global climate change, biotechnology, and energy production. BER's Microbial Genome Program helped shape microbial research and lead to BER's Genomic Science Program.

1.3. MICROBIAL GENOME PROGRAM (1994-2005)

Microbes have been found surviving and thriving in an amazing diversity of habitats, in extremes of heat, cold, radiation, pressure, salinity, and acidity, often where no other life forms could exist. Identifying and harnessing their unique capabilities, which have evolved over 3.8 billion years, will offer us new solutions to the longstanding challenges in environmental and waste cleanup, energy production and use, medicine, industrial processes, agriculture, and other areas. These wonderful capabilities of microbes will be soon added to the list of traditional commercial uses for microbes in the brewing, baking, dairy, and other industries.

1.3.1. Origin of microbial genome program

To explore the possibilities for new applications, in 1994 the U.S. Department of Energy (DOE) established the Microbial Genome Program (MGP) as a companion to its Human Genome Program (HGP). A principal goal of this spin off project is to determine the complete DNA sequence—the genome—of a number of nonpathogenic microbes that may be useful to DOE in carrying out its missions. The microbes chosen for genomic sequencing were selected with broad input from the scientific community. The microbial diversity of the program is an absolute treasure trove for biotechnology, ecology, evolution, and bioremediation.

The fervent task of scientists helped to reveal the genetic structure and genetic information of microbes. This information is being used by governmental, academic, medical, and industrial scientists. The number of possible applications of this information is staggering. Sequenced genomes provide us with a genetic "parts" list, the next challenge is to explore how these parts come together to form a functioning organism.

1.3.2. Organisms chosen for MGP studies

In 1995, the MGP's first full year, DOE funded four microbial genome sequencing projects focused on the bacterium *Mycoplasma genitalium* and three other microbes. Now fully characterized, the tiny *M. genitalium* genome thought to have the smallest genome of any known free-living bacterium provides a model for a minimal set of genes necessary for life. Its genome contains only 580,000 base pairs of DNA and yet encodes 470 genes. Future studies on this and other minimal genomes will help increase our understanding of more complex genomes.

Among the oldest life forms known, the Archaea make up one of three phylogenetic or evolutionary domains into which all life is classified. The other two are the Eukarya and the Bacteria. Archaea found thriving in extreme environments of heat and cold, acidity, pressure, and salinity are known as extremophiles ("extreme-loving" organisms). Comparisons of the genomes of organisms from all three domains give better understanding about the evolution of all living things. Descriptions of MGP-supported research on some microbes are as follow:

- *Methanococcus jannaschii* was among the first archaea chosen for sequencing. In 1996 its completed sequencing and analysis confirmed that the "tree of life" has three domains, a hypothesis first advanced nearly 20 years ago by Carl Woese (University of Illinois) but not given much credence at the time. The single celled *M. jannaschii* was isolated from a sample collected beneath more than 8000 feet of water at the base of a deep-sea thermal vent on the floor of the Pacific Ocean. The microbe lives without the sunlight, oxygen, and organic carbon important to most other forms of life and uses carbon dioxide, nitrogen, and hydrogen expelled from the thermal vent for its life functions. When the entire DNA sequence of *M. jannaschii* was determined, scientists found that about 65% of its potential gene sequences were not related to any gene previously discovered, representing an exciting area for future investigation.
- The archaeon *Archaeoglobus fulgidus* and the bacterium *Thermotoga maritima* have potential for practical applications in industry and government funded environmental remediation. Because they thrive in water temperatures above the boiling point of water, these organisms may provide heat stable enzymes for use in industrial processes. These processes could include conversion of wastes to useful chemicals. *A. fulgidus* has the added capability of surviving at the high pressures associated with deep oil wells, and *T. maritima* metabolizes simple and complex carbohydrates, including glucose, sucrose, starch, xylan, and cellulose. Cellulose and xylan are the most abundant biopolymers on Earth and, through their conversion to fuels such as ethanol, have major potential as sources of renewable energy. Comparisons of the genomic sequences of these two microbes will contribute to a greater understanding of evolutionary relationships as well as high-temperature protein function.
- The archaeon *Pyrobaculum aerophilum*, first isolated from a boiling marine vent, thrives at temperatures close to the maximum tolerated by living systems (113° C). Unlike most hyperthermophiles, *P. aerophilum* is able to withstand oxygen and can thus be manipulated more easily in the laboratory. Also, the proteins encoded by hyperthermophilic genomes are more stable than those of organisms living in more temperate environments.

- The bacterium *Shewanella putrefaciens*, which can grow with or without oxygen, is an excellent model system for manipulating organisms for bioremediation. Whole genome sequencing will elucidate metabolic pathways including those involved in corrosion, consumption of toxic organic pollutants, removal of toxic metals and radiation waste by conversion to insoluble forms. Other organisms that could be of great genetic and biochemical interest are present in extreme surface environments but are almost impossible to grow in the laboratory.
- *Deinococcus radiodurans*, an organism with extreme radiation resistance enables the microbe to survive doses thousands of times higher than that would kill most organisms, including humans. This bacterium was discovered in 1956 when it was identified as the culprit in a can of spoiled ground beef thought to be radiation sterilized. The remarkable DNA-repair processes of *D. radiodurans* allow it to stitch together flawlessly its own radiation-shattered genome in about 24 hours. Harnessing the potential of *D. radiodurans* in cleaning up waste sites containing radiation and toxic chemicals by exploring ways to add genes from other organisms to expand *D. radiodurans* capabilities for removing toxic wastes from contaminated sites. Its DNA sequence was completely determined in 1999. The added genes encode proteins that transform heavy metals to a more benign biomass and allow the concentration of heavy metals and the breakdown of organic solvents such as toluene. Studies into this organism's remarkable DNA-repair pathways also help to better understand how defects in human cellular processes might lead to the development of cancers.

These collections offer a rich resource for identifying and isolating novel species with potentially unique sets of genes as well as proteins with environmental, energy, biotechnological, and other applications.

1.4. MICROBIAL GENOME RESEARCH – PAST AND PRESENT

Microbial genome research has been funded through several different U.S. Department of Energy (DOE) Office of Science programs.

a) **The Microbial Cell Project (MCP):** The MCP was a 2000 initiative that was incorporated into the Genomic Science Program (GSP) Project in 2002. Microbial cells have internal organization and complex control systems that allow them to respond to their environment. They can work as miniature chemistry laboratories, making unique products and carrying out specialized functions. Ultimately, understanding the complex functioning of a single microbial cell will enable science to go far beyond just exploiting the beneficial capabilities of microbes to meet DOE's missions. The knowledge gained will apply to cells in all living things. Thus the MCP represents a first step in moving from cataloguing molecular parts to constructing an integrative view of life at the level of a whole organism — microbe, plant, or animal.

b) **Microbial Genome Program (MGP) [1994-2005]:** The MGP was begun in 1994 as a spinoff from the Human Genome Program. The goal of the program was to sequence the genomes of a number of nonpathogenic microbes that would be useful in solving DOE's mission challenges in environmental-waste cleanup, energy production, carbon cycling, and biotechnology. The Microbial Genome Program (MGP), initiated by DOE's Office of Biological and Environmental Research (BER) in 1994, has sequenced more than 485 microbial genomes and 30 microbial communities having specialized biological capabilities. Identifying these genes will help investigators discern how gene activities in whole living systems are orchestrated to solve myriad life challenges.

c) **Laboratory Science Program at the DOE Joint Genome Institute [2005-2008]:** Created in 2005, the goal of the Laboratory Science Program was to facilitate sequence-based science at the DOE National Laboratories, to develop cross laboratory large-scale sequencing projects that advance DOE missions, and to develop avenues to shape the service and product outputs of the Production Genomics Facility to meet the needs of the National Laboratories. The LSP allocated approximately ten billion bases (i.e., ten gigabases) of raw sequence per year.

d) **Community Sequencing Program at the DOE Joint Genome Institute:** The Community Sequencing Program (CSP) is a proposal-based program designed to provide access to high-throughput sequencing to the broadest possible community of researchers and to expand the diversity of disciplines using sequence data to address scientific questions. Proposals from researchers within the DOE National Laboratory system are accepted, and proposals from foreign investigators are particularly encouraged. Sequencing projects will be chosen based on scientific merit, judged through independent peer review and relevance to issues in global carbon cycling, alternative energy production, and bioremediation.

e) **Genome Biology Program (GBP):** The goal of the Genome Biology Program (GBP) is to understand the structure and function of various microorganisms and microbial communities and elucidate the evolutionary dynamics that shape their genomes. To accomplish that members of the program are developing pipelines for processing sequencing data, as well as methods and toolkits that facilitate their own genome analysis efforts and support those of the scientific community at large.

The ongoing projects of the Genome Biology program include (a) the development of large scale data integration systems for comparative analysis, annotation and interpretation of isolated genomes and metagenomes, such as the Integrated Microbial Genome systems (IMG and IMG/M) (b) the development and application of fully automated annotation pipelines (i.e. gene and function prediction) to isolate genomes and metagenomes (c) the development and application of pipelines for evaluating the quality of the automated gene predictions and their correction (Gene models Quality Control) (d) the development and curation of control vocabularies for the representation of gene function and their pathways which significantly improves the speed and accuracy of gene annotation and genome interpretation (e) the development and application of tools for processing and analyzing microbial communities sequenced with various technology platforms (i.e. both Sanger and 454 technologies), and others.

The Genome Biology Program also has a major research interest and focus on bioenergy with ongoing research activities on (a) archaeal genomics and methane production, (b) genomics of microorganisms and communities that decompose plant biomass and (c) bioengineering of lipid metabolism for biodiesel production.

f) **Genomic Science Program (GSP):** The U.S. Department of Energy's Genomic Science program (formerly Genomics:GTL) uses microbial and plant genomic data, high-throughput analytical technologies, and modeling and simulation to develop a predictive understanding of biological systems behavior relevant to solving energy and environmental challenges including bioenergy production, environmental remediation, and climate stabilization.

The Genomics: GTL (GTL) program, begun in 2001, is devoted to mechanistically understanding how microbes use a variety of energy sources, process metals, cycle carbon and nutrients such as nitrogen, and ameliorate radiation damage. Using completed genome sequences and a systems biology approach, GSP's goal is to develop a set of comprehensive principles and models showing how living systems function and to help put that knowledge to work in biotechnological solutions to DOE mission challenges.

Among other biological sciences, microbiology has been one of the greatest beneficiaries of the breakthrough in genomics and bioinformatics technologies that followed after the first whole prokaryotic genome sequence was published in July 1995 that of *Haemophilus influenzae* (Fraser *et al.*, 2002). Thus, to date, about 1800 bacterial genome assemblies have been "finished" at great expense with the aid of manual laboratory and computational processes that typically iterate over a period of months or even years (Genome Research,

2012). The recent short gun sequencing has revealed the size of the smallest genome as 2-6 Mb.

In the past decade, the progress in DNA sequencing and assembly, the faster generation of shotgun sequences, and the use of sophisticated methods for annotation have reduced the time required for each stage of a genome project and the cost per base pair, resulting in a finished product of higher quality. The improvements in sequencing have been accompanied by free access to these sequences in public databases. These public databases can aid scientists in isolating genes, comparing genomes, relating species evolutionarily, and speculating on the presence and function of genes, and consequently of the proteins that genes code for. The information derived from whole genome sequences following their comparative analysis can be used in studies that search for novel aspects of biochemistry, physiology, and metabolism of these organisms to investigate the roles microorganisms play in complex ecosystems and in global geochemical cycles, to study their diversity, to predict the impact microorganisms have on the productivity and sustainability of agriculture and forestry and on the safety and quality of food supply. Also, new genome sequences can be used to infer phylogenetic relationships among prokaryotes that deal with the organization and evolution of microbial genomes, mechanisms of transmission, exchange and reshuffling of genetic information.

Some commonly used public sequence databases are:

1. GOLD ™ genomes online database
2. The Institute for Genomic Research - Microbial database(TIGR)
3. National Centre for Biotechnology Information - Microbial genome (NCBI)
4. The Welcome Trust Sanger Institute – Microbial genome
5. Microbial Genome Database (MGBD) 2013 –microbial genome database for comparative analysis

Given that many organisms will not be coaxed readily into pure culture, a critical advance is to extend the understanding of the uncultured world beyond cataloging 16S rRNA gene sequences. Microbiologists have striven to devise methods to analyze the physiology and ecology of these diverse, uncultured organisms. This gave birth to an exciting new field called 'metagenomics' to explore the unseen microbial world.

1.5. METAGENOMICS- THE SCIENCE OF GENETIC DIVERSITY

Metagenomics is the culture-independent genomic analysis of microbial communities. The term is derived from the statistical concept of meta-analysis (the process of statistically combining separate analyses) and genomics (the comprehensive analysis of an organism's genetic material) (Rondon *et al.*, 2000). Metagenomics can be used to address the challenge of studying prokaryotes in the environment that are, as yet, unculturable and which represent more than 99% of the organisms in some environments. This approach builds on recent advances in microbial genomics and in the polymerase chain reaction (PCR) amplification and cloning of genes that share sequence similarity (e.g. 16S rRNA, *nif*, *recA*) directly from environmental samples. Whereas PCR amplification requires prior knowledge of the sequence of the gene to design primers for amplification but direct cloning of DNA for genomics can theoretically access genes of any sequence or function. Direct genomic cloning offers the opportunity to capture operons or genes encoding pathways that may direct the synthesis of complex molecules, such as antibiotics. Sequence information about the genes flanking a gene of particular interest can also be obtained, potentially providing insight into the genomic environment of the gene or the phylogenetic affiliation of the organism from which it was derived. Moreover, a long-term goal of metagenomic analysis is to reconstruct the genomes of uncultured organisms by identifying overlapping fragments in metagenomic libraries and 'walking', clone to clone, to assemble each chromosome.

1.6. WHY METAGENOMICS

Microbes run the world. *"Microbes can live without human but humans cannot live without microbes"*. God has empowered the microbial world with that much potentialities and power to sustain our universe. Although we can't usually see them, microbes are essential for every part of human life indeed all life on Earth. Every process in the biosphere is touched by the seemingly endless capacity of microbes to transform the world around them. The chemical cycles that converts the key elements of life carbon, nitrogen, oxygen, and sulfur into biologically accessible forms are largely directed by and dependent on microbes. All plants and animals have closely associated microbial communities that make necessary nutrients, metals, and vitamins available to their hosts. Through fermentation and other natural processes, microbes create or add value to many foods that are staples of the human diet. We depend on microbes to remediate toxins in the environment both the ones that are produced naturally and the ones that are the byproducts of human activities, such as oil and chemical spills. The microbes associated with the human body in the intestine and mouth enable us to extract energy from food that we could not digest without them and protect us against disease-causing agents.

These functions are conducted within complex microbial communities intricate, balanced and integrated entities that adapt swiftly and flexibly to environmental change. Metagenomics is a new tool to study microbes in the complex communities where they live and to begin to understand how these communities work. Traditional microbiological approaches have already shown how useful microbes can be. To circumvent some of the limitations of cultivation approaches, indirect molecular methods based on the isolation and analysis of nucleic acids (mainly DNA) from soil samples without cultivation of microorganisms have been developed. The new approach of metagenomics will greatly extend our ability to discover and derive benefits from microbial world.

Mini Quiz

1. What is the significance of rRNA analysis in metagenomics?
2. Elucidate the microbial genome programme and the microbes supported by it?
3. Mention few genes that share sequence similarity?
4. Metagenomics approach is superior to traditional microbiological approach. Substantiate?

CHAPTER - 2

SOIL METAGENOMICS – THE CULTURE INDEPENDENT INSIGHT

The word metagenomics was coined by Handelsman (1998) to capture the notion of analysis of a collection of similar but not identical items, as in a meta-analysis, which is an analysis of analyses. Community genomics, environmental genomics, and population genomics are synonyms for the same approach. The phylogenetic and functional diversity of microorganisms in various habitats, including soil, vastly exceeds the diversity of prokaryotic phyla known from cultivation. The soil environment is an abundant yet under-characterized source of genetic diversity that has great potential to enrich our understanding of soil microbial ecology and provide enzymes and bioactive compounds useful to human society. Fortunately, the recent development of metagenomic and other culture-independent approaches has enabled investigation of the functional genetic diversity of soil microorganisms without the inherent biases of cultivation.

2.1. WHAT IS METAGENOMICS?

Metagenomics can be defined as the genomic analysis of the collective microbial assemblage found in an environmental sample (Handelsman *et al.*, 1998). There are many variants on metagenomic approaches, which initially were dependent upon cloning of DNA from an environmental sample but more recently many metagenomic approaches relies upon high-throughput sequencing. One of the main advantages of functional metagenomics is its ability to identify gene products from as-yet-uncultured microbes with no significant homolog within the GenBank database.

Metagenomic analysis involves isolating DNA from an environmental sample, cloning the DNA into a suitable vector, transforming the clones into a host bacterium, and screening the resulting transformants (Fig. 2.1). The clones can be screened for phylogenetic markers or "anchors," such as 16S rRNA and *recA*, or for other conserved genes by hybridization or multiplex PCR or for expression of specific traits, such as enzyme activity or antibiotic production or they can be sequenced randomly. Each approach has strengths and limitations; together these approaches have enriched our understanding of the uncultured world, providing insight into groups of prokaryotes that are otherwise entirely unknown. The approaches to study the soil metagenome can be broadly classified into two.

a) Sequence driven analysis
b) Function driven analysis

2.2. SEQUENCE DRIVEN ANALYSIS

a) **Detection with hybridization probes:** Sequence-driven analysis relies on the use of conserved DNA sequences to design hybridization probes or PCR primers to screen metagenomic libraries for clones that contain sequences of interest. It also involves complete sequencing of clones containing phylogenetic anchors that indicate the taxonomic group. Alternatively, random sequencing can be conducted, and once a gene of interest is identified, phylogenetic anchors can be sought in the flanking DNA to provide a link of phylogeny. Sequence analysis guided by the identification of phylogenetic markers is a powerful approach first proposed by DeLong and his co-workers. They produced the first genomic sequence linked to 16S rRNA gene of an uncultured archaeon. The sequence of flanking DNA revealed a bacterio-rhodopsin like gene. Its gene product was authentic photoreceptor, leading to the insight that bacteriorhodopsin genes are not limited to *Archaea* but is abundant among the *Proteobacteria* of the ocean.

b) **Sequencing random clones:** The alternative to a phylogenetic marker-driven approach is to sequence random clones, which has produced dramatic insights, especially when conducted on a massive scale. The distribution and redundancy of functions in a community, linkage of traits, genomic organization, and horizontal gene transfer can be inferred from sequence-based analysis

   The application of sequence-based approaches involves the design of DNA probes or primers which are derived from conserved regions of already-known genes or protein families. In this way, only novel variants of known functional classes of proteins can be identified. Nevertheless, this strategy has led to the successful identification of genes encoding novel enzymes, such as dimethyl sulfonio propionate degrading enzymes, dioxygenases, nitrite reductases, [Fe-Fe] hydrogenases, [Ni-Fe] hydrogenases, hydrazine oxidoreductases,chitinases, and glycerol dehydratases.

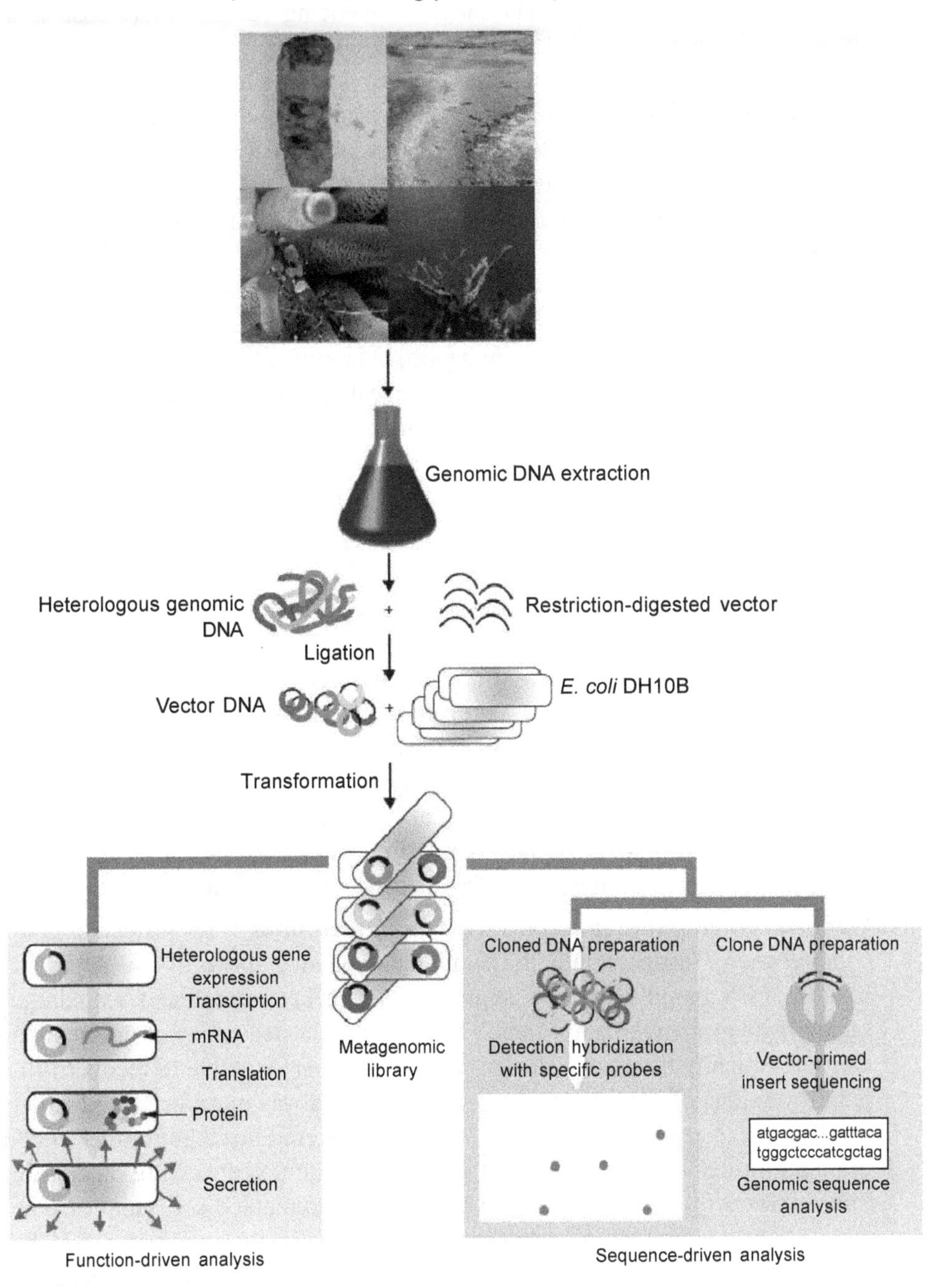

Fig. 2.1. Schematic representation of metagenomic analysis

c) **Pyrosequencing and other next-generation approaches:** Pyrosequencing and other next-generation approaches offer the capacity for massively parallel sequencing of metagenomic samples. The accuracy of pyrosequencing is comparable to that achieved via Sanger sequencing but it is more cost- and time-effective per sequenced nucleotide and sequencing read length has been gradually increasing with each iteration of sequencing technologies. The increased availability of high-throughput sequencing technologies has made it possible to gain access to the genetic diversity within environmental communities. Pyrosequencing has been used in the investigations of microbial diversity in soil deep sea ecosystems and phage populations from various environments. Because pyrosequencing relies on an amplification process, the same environmental contamination challenges that apply to PCR-based applications also apply to pyrosequencing. However, since pyrosequencing currently generates reads only 300-500 bp in length, obtaining intact, larger DNA is not critical. Alternatively, the DNA template can be diluted (along with contaminants) to permit PCR amplification or bovine serum albumin can be added to the reaction mixture to prevent humic acid-mediated inhibition. Pyrosequencing and next generation technologies are dealt separately in chapter 6.

Read length is a critical factor in the probability that a metagenomic sequence finds a significant hit within GenBank or other database. Even for a pure bacterial culture, it is not uncommon for a completely sequenced bacterial genome to have 35% to 45% of predicted open reading frames (ORFs) with no significant homolog in GenBank .This problem is only exacerbated with metagenomic sequences, with an even larger proportion of metagenomic sequences from soil and other environments having no significant BLAST homolog. Even with the difficulty in interpretation of much of the sequences within metagenomic datasets, substantial information related to the genomic composition, and predicted functions and metabolic pathways, of microbial communities has been unearthed from deep-sequencing approaches.

d) **Gene-targeted metagenomics (GT-metagenomics):** In order to gain comprehensive insights into the available sequence space of the genes of interest, PCR-based screening approaches have been combined with large scale pyrosequencing of amplicons. The thereby-collected sequence information can subsequently be used to design probes which are suitable to recover full-length versions of the target genes. This approach was introduced by Iwai *et al.* (2010). The authors termed it gene-targeted metagenomics (GT-metagenomics). It was applied to recover genes encoding aromatic dioxygenases from polychlorinated-biphenyl-contaminated soil samples. The authors employed a PCR primer set that was directed against a 524-bp conserved region which confers substrate specificity to biphenyl dioxygenases.

2.3. FUNCTION-DRIVEN ANALYSIS

The function-driven analysis is initiated by identification of clones that express a desired trait, followed by characterization of the active clones by sequence and biochemical analysis. This approach quickly identifies clones that have potential applications in medicine, agriculture or industry by focusing on natural products or proteins that have useful activities. The limitations of the approach are that it requires expression of the function of interest in the host cell and clustering of all of the genes required for the function. It also depends on the availability of an assay for the function of interest that can be performed efficiently on vast libraries, because the frequency of active clones is quite low. Many approaches are being developed to mitigate these limitations. They are:

i) Heterologous expression
ii) Identifying active clones—screens, selections, and functional anchors
iii) Random sequencing of clones
iv) Microarray
v) Direct testing of colonies for specific function
vi) Heterologous complementation

i) **Heterologous expression:** A powerful yet challenging approach to metagenomic analysis is to identify clones that express a function. Success requires faithful transcription and translation of the gene or genes of interest and secretion of the gene product, if it is extracellular. This approach has the potential to identify entirely new classes of genes for new or known functions. The significant limitation is that many genes, perhaps most, will not be expressed in any particular host bacterium selected for cloning. In fact, there is an inherent contradiction in this approach genes are cloned from exotic organisms to discover new motifs in biology, and yet these genes are required to be expressed in *Escherichia coli* or another domesticated bacterium in order to be detected. The DNA of diverse organisms is successfully expressed in *E. coli* but heterologous expression remains a barrier to extract the maximum information from functional metagenomics analyses.

Improved systems for heterologous gene expression are being developed with shuttle vectors that facilitate screening of the metagenomic DNA in diverse host species and with modifications of *Escherichia coli* to expand the range of gene expression. Although the genes encoding the enzymes required for synthesis of secondary metabolites are usually clustered on a contiguous piece of DNA, obtaining fragments of DNA large enough to contain the information required for synthesis of complex antibiotics, which can require over 100 genes, presents a challenge. This is a particular problem when the DNA is isolated from soil or other environments that contain high concentrations of contaminants that inhibit cloning. The choice of vectors and construction of expression libraries are detailed in chapter 6.

ii) **Identifying active clones screens, selections, and functional anchors:** The frequency of metagenomic clones that express any given activity is low. To address the challenge of detecting rare, active clones in large libraries, efforts are being directed toward the design of highly sensitive assays and robotic screens that efficiently detect low levels of activity in many samples. The most convenient traits to study are those that present a selectable phenotype, such as resistance to an antibiotic or growth on an unusual substrate, because selections are orders of magnitude more efficient than screens.

For instance, in a search for lipolytic clones derived from German soil, only 1 in 730,000 clones showed activity (Henne *et al.*, 2000). In a library of DNA from North American soil, 29 of a total of 25,000 clones expressed hemolytic activity (Rondon *et al.*, 2000). The scarcity of active clones therefore necessitates development of efficient screens and selections for discovery of new activities or molecules. Just as bacterial genetics relies on selections to detect low-frequency events, metagenomics will be advanced by seeking selectable phenotypes to increase the collection of active clones that can be compared, analyzed, and used to build a conceptual framework for functional analysis. Genes that are expressed in an ordinary host such as *E. coli* may be extraordinary and novel.

High-throughput screens can substitute when the functions of interest do not provide the basis for selection. For example, on certain indicator media, active clones display a characteristic and easily distinguishable appearance even when plated at high density. With the indicator dye tetrazolium chloride, Henne *et al.* (1999) detected clones that utilize 4-hydroxybutyrate in libraries of DNA from agricultural or river valley soil. Very rare lipolytic clones in the same libraries were detected by production of clear halos on media containing rhodamine and either triolein or tributryin.

The discovery of new biological motifs will depend in part on functional analysis of metagenomic clones. Functional screens of metagenomic libraries have led to the assignment of functions to numerous "hypothetical proteins" in the databases. Innovation will be required to identify and overcome the barriers to heterologous gene expression and to detect rare clones efficiently in the immense libraries that are needed to represent all of the genomes in complex environments, such as soil.

An emerging and powerful direction for metagenomic analysis is the use of *functional anchors*, which are the functional analogs of phylogenetic anchors. Functional anchors are functions that can be assessed rapidly in all of the clones in a library. When a collection of clones with a common function is assembled, they can be sequenced to find phylogenetic anchors and genomic structure in the flanking DNA. Such an analysis can provide a slice of the metagenome that cuts across clones with a different selective tool, determining the diversity of genomes that contain a particular function that can be expressed in the host carrying the library. Technological developments that promote functional expression and screening will advance this new frontier of functional genomics.

Functional screens of metagenomic libraries have identified both novel and previously described antibiotics an antibiotic resistance gene, lipases, chitinases, membrane proteins, genes encoding enzymes for the metabolism of 4-hydroxybutyrate and genes encoding the biotin synthetic pathways. Functional analysis has identified novel antibiotics, antibiotic resistance genes,$Na^+(Li^+)/H^+$ transporters, and degradative enzymes.

iii) **Random sequencing of clones:** Theoretically, of soil-derived libraries is another approach to characterize the soil ecosystem on a genomic level, but the species-richness of soil habitats would require enormous sequencing and assembly efforts.

iv) **Microarray:** Microarray technology could be useful for analyzing the soil metagenome and profiling metagenomic libraries. For example, genes encoding key reactions in the nitrogen cycle were detected using microarrays from samples that were collected from soil, and provided information on the composition and activity of the complex soil microbial community. However, microarray methods for gene detection show a 100 to 10,000-fold lower sensitivity than PCR. This difference might prevent the analysis of sequences from low-abundance soil microorganisms. The improvement of sensitivity and specificity are among the challenges of using complex soil DNA or RNA with microarray technology.

v) **Direct testing of colonies for specific function:** Function-driven approaches can include the direct testing of colonies for a specific function. For example, chemical dyes and insoluble or chromophore-bearing derivatives of enzyme substrates can be incorporated into the growth medium solidified with agar to monitor enzymatic functions of individual clones. The sensitivity of these screens makes it possible to detect rare clones. An example is the screening of soil-based libraries for genes conferring polyol oxidoreductase activity, which was based on the ability of the recombinant *E. coli* strains to form carbonyls from polyols (Fig. 2.2a). Another example is the detection of *E. coli* clones with proteolytic activity on agar plates containing skimmed milk (Fig.2. 2b).

vi) Heterologous complementation: Another approach that allows detection of functional clones is the use of host strains or mutants of host strains that require heterologous complementation for growth under selective conditions. An example is complementation of a Na+/H+ antiporter-deficient *E. coli* strain with soil-derived libraries, which led to the identification of two new genes that encode Na+/H+ antiporters from a soil library consisting of 1,480,000 clones.

2.3.1. Limitations of function-driven screening

i) Most of the biomolecules recovered by function-driven screens of complex soil libraries are either weakly related or entirely unrelated to known genes, and rediscovery of genes has not been reported. This confirmed that the amount of soil DNA that has been cloned and screened only represents the tip of the iceberg with respect to discovery of new natural products from the soil metagenome.

ii) The frequency of soil-derived metagenomic clones that express a specific activity is usually low, so large numbers of clones have to be tested.

iii) Although function-driven screens usually result in identification of full-length genes (and therefore functional gene products), its reliance on the expression of the cloned gene(s) and the functioning of the encoded protein in a foreign host is limited.

a) Detection of clones harbouring genes that confer carbonyl formation. Screening is based on the ability of the library-containing *Escherichia coli* clones to form carbonyls from test substrates, that is, polyols , during growth on indicator agar. The test substrates are included in the indicator agar, which contains a mixture of pararosaniline and sodium bisulphite (Schiff reagent). The production of carbonyls from test substrates on indicator plates by clones results in formation of a dark red Schiff base.The carbonyl-forming colonies are red and are surrounded by a red zone, whereas colonies failing to form carbonyls from the test substrate remain uncoloured.

b) Detection of proteolytic activity. Proteolytic *E. coli* clones are detected on agar media containing skimmed milk by zones of clearance around the colonies.

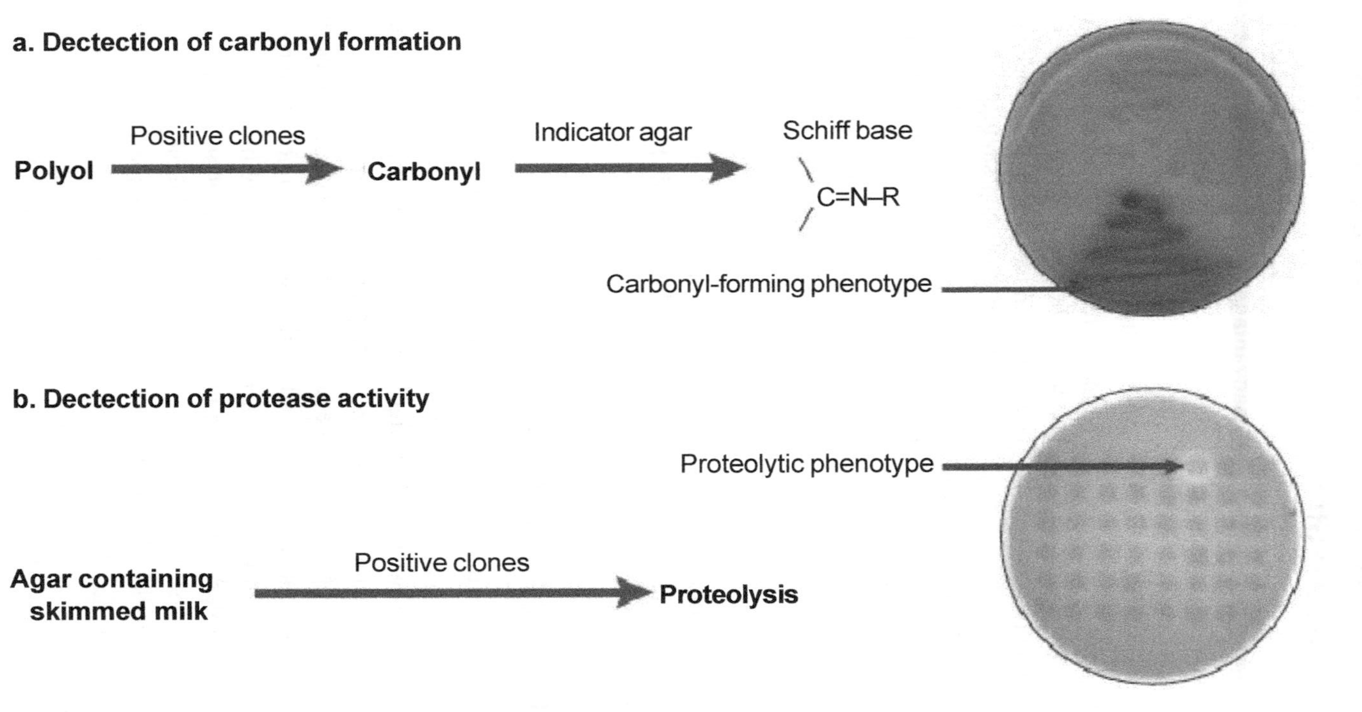

Fig. 2.2. Examples of activity-based screens.

2.4. FUNCTION-DRIVEN VERSUS SEQUENCE-DRIVEN SCREENING STRATEGIES

Advantages	Disadvantages
Function-driven screening method	
Completely novel genes can be recovered	Dependent on expression of the cloned genes by the bacterial host
Selects for full-length genes	Requires production of a functional gene product by the bacterial host
Selects for functional gene products	Dependent on the design of a simple activity-based screening strategy
Sequence-driven screening method	
Independent of expression of the cloned genes by the bacterial host used	Recovered genes are related to known genes
Similar screening strategies can be used for different targets, for example, colony hybridization and PCR	Partial genes can be cloned Not selective for functional gene products

Both the approaches require high quality metagenomic DNA. The techniques involved in accessing the soil metagenome and its further analysis has been dealt in the ensuing chapters.

2.5. APPLICATIONS OF METAGENOMICS

Many microorganisms have the ability to degrade waste products, make new drugs for medicine, make environmentally friendly plastics, or make even some of the ingredients of the food we eat. By isolating the DNA from these organisms, it provides us with the opportunity to optimize these processes and adapt them for use in our society. As a result of ineffective standard laboratory techniques, the potential of microbes is relatively untapped, unknown and uncharacterized. Metagenomics represents a powerful tool to access the abounding biodiversity of native environmental samples. The valuable property of metagenomics is that it provides the capacity to effectively characterize the genetic diversity present in samples regardless of the availability of laboratory culturing techniques. Information from metagenomic libraries has the ability to enrich the knowledge and applications of many aspects of industry, therapeutics, and environmental sustainability. This information can then be applied to society in an effort to create a healthy human population that lives in balance with the environment. The applications of metagenomics are briefed as follow:

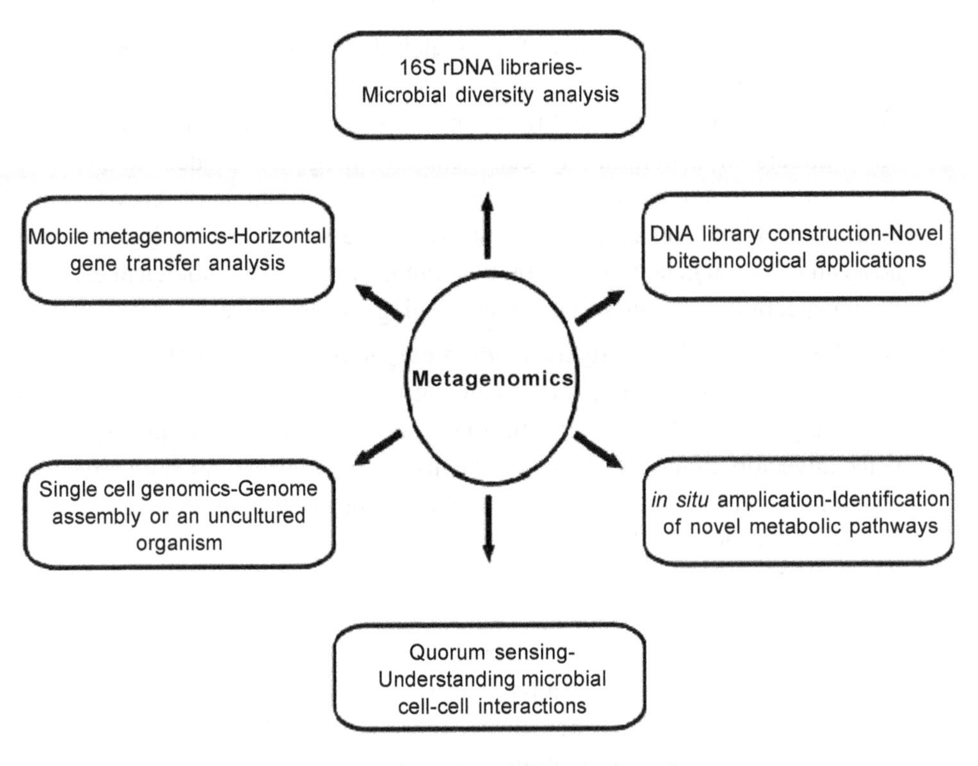

Fig. 2.3. Applications of Metagenomics

1. **Metagenomics as a tool for exploring soil microbial diversity:** The microorganisms classified in the prokaryotic domains of the tree of life viz., bacteria and archae possess immense metabolic diversity and their activities are critical in processes ranging from sewage treatment to regulating the composition of atmosphere. Especially in the light of the rate of modern climate change it is essential to understand how human activities such as agriculture, waste management and climate modification affect microbial communities. Thus discovering and understanding the diversity of microbial communities is a high priority in ecology. Sequence analysis guided by the identification of phylogenetic markers produced valuable information on bacterioprhodopsin gene in a proteobacteria from sea water which is a breakthrough in microbial diversity studies. So far it was believed that bacteriorhodopsin gene is an authentic photoreceptor of archaebacteria only.
2. **Ecological inference from metagenomics:** The exigent questions in microbial ecology focus on how microorganisms form symbioses with eukaryotes, compete and communicate with other microorganisms, acquire nutrients and produce energy. Thus far, metagenomics has provided insights into each of these areas, but in each instance, the challenge is to link the genomic information

with the organism or ecosystem from which the DNA is isolated. An exciting potential of metagenomics is to provide community wide assessment of metabolic and biogeochemical function. Analysis of specific functions across all members of a community can generate integrated models about how organisms share the workload of maintaining the nutrient and energy budgets of the community eg) acid mine drainage. Expression of a gene in a cultured host can establish gene function, but without the appropriate biological context, circumspection is required in drawing ecological inferences.

3. **Novel industrial biocatalysts from metagenomics:** Biocatalysis, the use of microbial cells or isolated enzyemes in th e production of fine chemicals is steadily moving towards and becoming accepted as an indespensible tool in the iinventory of modern synthetic chemistry. Application of modern screening technology to diversity is clearly a rewarding approach in the development of "white biotechnology" field. The metagenome approach will revolutionize the field of enzyme discovery by providing access to the unseen majority of microbial diversity and its enzymatic constituents. DNA analysis of cosmid libraries has identified several biocatalysts encoding genes such as amidase, cellulases, alpha amylases,1,4 alphaglucan branching enzymes, pectate lyases, lipases and DNAases. This opened new door for prospecting novel biocatalysts from metagenomes of any complex environment.
4. **Innovation of novel antibiotics by metagenomics:** Small molecule discovery by functional genomics has concentrated on antibiotics, which are of interest for their pharmaceutical applications as well as for their roles in ecosystem function. Traditional antibiotic screens have not been a rich source of novel antibiotics, because of the experimental limitations. With standard inhibition assays a Mycobacterium inhibiting antibiotic, terragine was discovered from a soil metagenome cloned in *Streptomyces lividans*.
5. **Metagenomics in quorum sensing regulation studies:** A sub-inhibitory concentration of many antibiotics induce quorum sensing despite no resemblance in structure to acylated homoserine lactones that appear to be natural inducers. This opportunity was investigated by designing a high- throughput screen to identify compounds that induce the expression of genes under the control of a quorum sensing promoter. Quorum sensing (QS)-mediated bacterial responses to cell density are specific to each bacterial species, and are important in understanding bacterial pathogenesis and other bacterial phenotypes in natural environments (e.g., bioluminescence of *Vibrio fischeri* within the light organ of the *Euprymna* squid). The use of a metagenomic approach to study QS regulation in the soil environment was pioneered by Williamson *et al.,(*2005) wherein they identified clones producing unknown molecules that activated

QS-regulated genes. Clones of interest were identified using a high throughput intracellular screen, i.e. the metagenomic DNA is within a host cell that contains a biosensor responsive to compounds inducing QS. Some of the QS molecules identified by metagenomics are:

- N-acylhomoserine lactone (NAHL) from clones obtained from pasture soil (Riaz *et al.*, 2008).
- Novel lactonase family proteins interfering with QS when expressed in *Pseudomonas aeruginosa* successfully inhibited motility and biofilm formation (Schipper *et al.*, 2008)
- Novel QS synthase genes from metagenomic libraries constructed with DNA isolated from activated sludge

6. **Role of metagenomics in bioremediation:** Xenobiotics include aromatic compounds and their derivatives, and polychlorinated biphenyls (PCBs), anthropogenic chemical pollutants that persist in the environment and are recalcitrant to complete removal. Xenobiotic degradation can be achieved by biotic and/or abiotic reactions, and may be accelerated by harnessing microbial degradative activities to biostimulate or bioaugment the natural attenuation of environmental contaminants. The application of metagenomics may aid in the isolation of novel catabolic pathways for degradation of xenobiotic compounds, indicating the functional genetic capacity for contaminant degradation and providing molecular tools useful for identification of the microbial taxa encoding the biodegradative gene(s). A combined approach using metagenomics and other molecular techniques is commonly used to study microorganisms useful for bioremediation of environmental contaminants. Labeled substrates have been used to target and recover genes from populations involved in the degradation process. [^{13}C]-labeled biphenyl was used to identify biphenyl dioxygenase genes from bacteria capable of growing in PCB-contaminated river sediments. Other metagenomic studies have identified catabolic pathways that encode nitrilases, which play an important role in both biosynthetic and catabolic reactions as well as enzymes with catalytic properties that degrade organic contaminants

   The strategies to improve the isolation of biosynthetic and catabolic pathways are:

   i) to enrich the environmental sample for microbial populations capable of growth on certain substrates or for survival under different physico-chemical conditions.

   ii) stable isotope probing (SIP) is an approach that enriches the DNA (or RNA) of microorganisms that can utilize a stable isotope (e.g., ^{13}C-glucose) and incorporate the isotope into newly synthesized nucleic

iii) substrate-induced gene expression screening (SIGEX). SIGEX identifies clones from an operon-trap metagenomic library that are induced in the presence of a specific substrate, resulting in green fluorescence protein expression that can be detected using fluorescence activated cell sorting.

7. **Population Genetics and Microheterogeneity:** Metagenomic analysis has revealed that even apparently uniform populations contain substantial microheterogeneity. If genetic variation in the environmental population is of interest, it may be more productive to clone the genome from the natural population than analyze the genomes of individuals cultured from it. Within the population of *Cenarchaeum symbiosum* associated with the marine sponge, the rRNA genes are highly conserved, showing >99.2% identity, which indicates that the population comprises a single species. In the genomic regions flanking the rRNA genes, however the DNA sequence identity drops to 87.8%. A high frequency of single-nucleotide polymorphisms was observed among the strains of the same species. The Ferroplasma type II group appears to contain a composite genome, with segments derived from three sources. In contrast, the Leptospirillum group II genome contained very few single nucleotide or large-scale genome polymorphisms.

Functional and sequence-based screening of soil-based libraries has provided insights into soil microbial communities and has led to the identification of novel biomolecules, but these approaches have strengths and limitations (Table 2.4). To take full advantage of the enormous diversity of soil microorganisms, a combination of sequence-based and functional approaches and of different types of libraries should be used to probe the soil metagenome.

Mini Quiz

1. Define soil metagenome? Explain the approaches involved in decoding soil metagenome?
2. What is pyrosequencing? Narrate its merits over Sanger sequencing?
3. Distinguish sequence and function driven sequencing strategies?
4. Elaborate the role of metagenomics in quorum sensing with suitable example?
5. What is population genetics?

CHAPTER - 3

TECHNIQUES IN SOIL METAGENOMICS

Even before the advent of genome sequencing projects and the subsequent development of transcriptomic, proteomic and metabolomic tools the potential to assess the status of environmentally relevant organisms through measurement of their genes and proteins was already being investigated. The type of assay to which genomic approaches could be applied spanned the organisation cascade. Many of the methods and studies focused on assessing changes in the expression of single genes and also organelle level changes. Some, however, have exploited molecular methods to monitor change at the community level in both bacterial and eukaryotic species. The majority of these techniques require a considerable level of specialised knowledge (technical and theoretical), however, a number are being developed to provide solutions targeted at the monitoring market. In the development of all of these molecular based assays, the most significant advances came upon the development of the polymerase chain reaction (PCR).

PCR was devised by Mullis *et al.* (1986) and has proved itself to be the most versatile yet precise of all the biological techniques. During the PCR reaction, specific DNA sequences are multiplied by sequential splitting of the double stranded DNA molecule, annealing of specially designed complimentary oligonucleotide primers and extension to form a complimentary strand under the action of heat stable DNA polymerase. A series of heating and cooling cycles are used to drive the splitting, annealing and extension phases. To date PCR based methods have been applied in a number of techniques with potential for biological assessment of environmental quality.

3.1. MICROBIAL COMMUNITY PROFILING

A number of techniques suitable for the analysis of the composition of the community are available and these have been widely used in many research projects. These can be either culture dependent or culture independent. Culture independent studies extract total DNA or RNA from microbial communities and use universal forward and reverse primers in combination with the PCR reaction to amplify species or genera specific (usually the 16S ribosomal subunit) DNA fragments from a whole community sample previously isolated directly from soil. After PCR, fragments can be used in a range of post amplification analyses. Two quantitative "gold standard" techniques exist that are non-invasive i.e. they do not rely on the extract of DNA from microbial cells and thus can be used to examine microbial ecology in situ. These are flow cytometry and Fluorescence in situ hybridisation (FISH).

3.1.1. Denaturing and temperature gradient gel electrophoresis (DGGE and TGGE)

PCR products are separated by electrophoresis on an acrylamide gel containing a denaturing urea gradient (DGGE) or a temperature gradient (TGGE). The denaturing conditions induce strand melting at a point dependent on the nucleotide composition. These melted fragments migrate slower through the gel matrix, thus separating the fragments of differing nucleotide composition.

3.1.2. Single strand confirmation polymorphism (SSCP)

PCR is conducted with a phosphorylated and non-phosphorylated primer. Products are converted to single strands by lambda exonuclease digestion of the phosphorylated strand. These single strands are electrophoresed on a non-denaturing gel where they separate according to sequence-dependent folding confirmations that affect their mobility.

3.1.3. Amplified ribosomal DNA restriction analysis (ARDRA)

PCR amplification is followed by cutting with restriction enzymes. Digests are then electrophoresed on agarose or acrylamide gels allowing identification of sequence dependent banding patterns. One methodological problem with ARDRA for quantitative estimation of microbial diversity is that the number of bands in the profile is always greater than the number of amplified PCR products. This technique is usually used to analyse clone libraries to screen for identical clones sets in order to prioritise and reduce redundancy in subsequent DNA sequencing efforts.

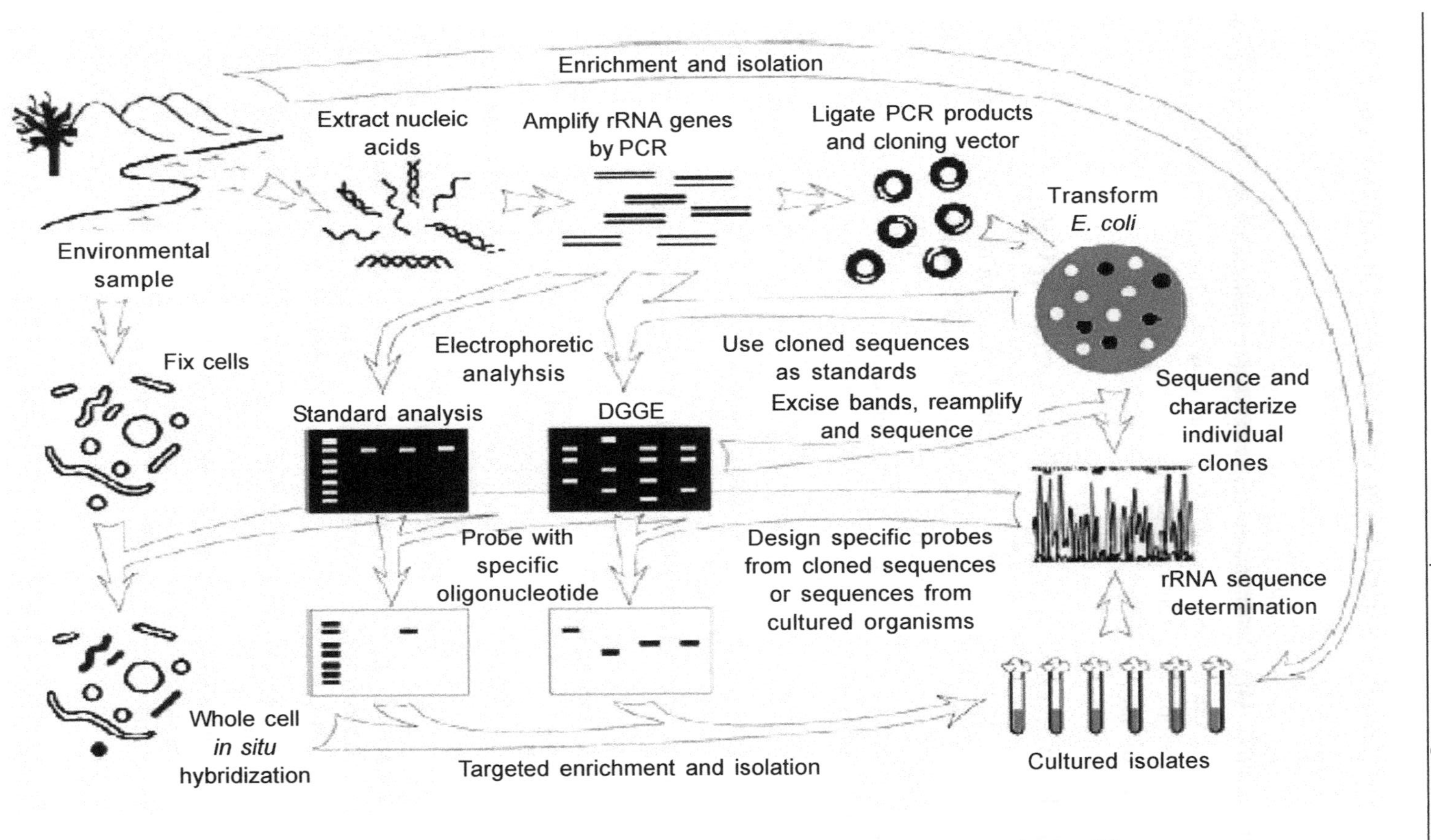

Fig. 3.1. Schematic representation of approaches to microbial community profiling

3.1.4. Terminal-restriction fragment length polymorphism (T-RFLP)

T-RFLP is similar to ARDRA, but a fluorescent primer is included in the PCR reagent mixture. After restriction enzyme digestion, fragments are analysed on an automated sequencer. Only those bands carrying the fluorescent label (*i.e.* the terminal restriction fragment) are detected. This number corresponds directly to the number of species.

3.1.5. Fluorescence *in situ* hybridisation (FISH)

FISH allows the direct scrutiny of microbial populations within their three-dimensional ecological niche. An environmental sample is fixed, using paraformaldehyde and the cell membranes permeabilised. Fluorescent oligonucleotide probes are introduced into the samples that are specific for certain genera or species of bacteria. Under certain conditions these probes will only hybridise with the specific bacteria against which they are targeted towards. After washing excess probe and filtration, samples can be visualised by either epifluorescence or confocal laser-scanning microscopy (CLSM; Figure 3.2). CLSM is particularly useful for samples such as activated sludge where auto-fluorescence is a problem, as the laser will only excite the sample in a single focal plane.

3.1.6. Flow cytometry

Flow cytometry is a generic technology which counts and measures multiple characteristics of individual particles in a flow stream. The initial application of flow cytometry in aquatic science was the analysis of phytoplankton, since the red chlorophyll fluorescence from chlorophyll can be used for the detection of photosynthetic cells even in the presence of large quantities of detritus. This is now being used as a tool to identify different taxonomic groups of bacteria and phytoplankton using spectral differences in auto fluorescence excitation and emission. Flow cytometry can also be coupled with the use of fluorescent oligonucleotide probes to detect and quantify specific species or genera of microorganisms in complex environmental mixtures. Flow cytometry is an inherently fast technique. Thousands of cells can be analysed per second. Routinely, 10,000 cells are analysed in one sample and up to 30 samples can be analysed per hour. Moreover, particles of interest can be physically sorted for subsequent analysis or purification of cultures.

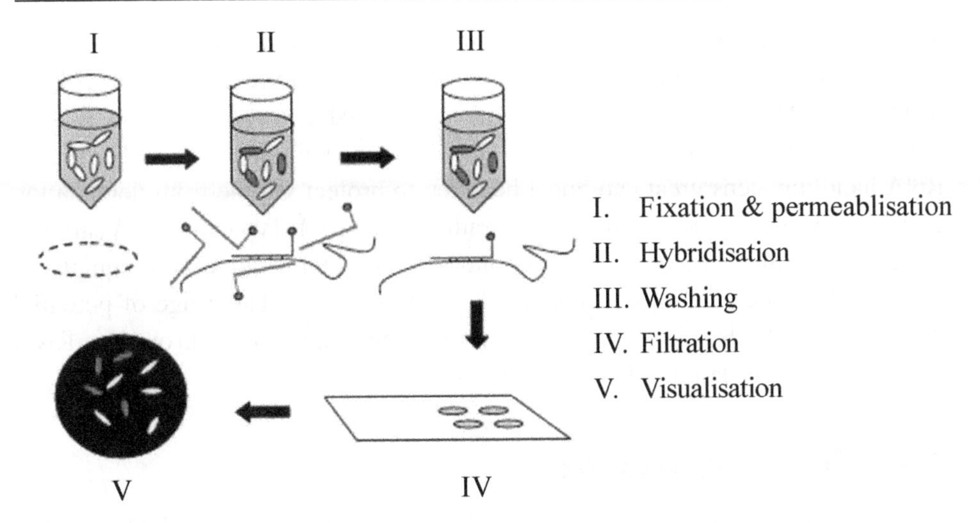

Fig. 3.2. Fluorescence in situ hybridisation (FISH) of microbial populations

3.2. EUKARYOTIC COMMUNITY PROFILING

Identification of the presence of eukaryotic species has traditionally relied on morphologically based taxonomy, a task that is often laborious. As an answer to this problem, it has been suggested that PCR based 'molecular barcoding' techniques could be used. The method is based on the amplification and sequencing of specific regions of the small subunit ribosomal (SSU) RNA gene. Identified sequences may then be categorised into molecular taxonomic units based on sequence similarity and bioinformatic search of known SSU sequences. The approach allows the presence of specific genera or species to be identified within a community and can be linked either to classical biodiversity statistics, specific indices (such as the nematode maturity index) or potentially to multivariate based analytical tools such as the RIVPACS system. To date the molecular barcoding has been developed for nematodes in soil systems (Floyd *et al.*, 2002). Other groups that may be suitable include oribatid mites and collembolans in soil and chironomids in freshwaters.

3.3. SINGLE GENE TRANSCRIPT QUANTIFICATION METHODS

3.3.1. Sample preparation

Quantitation of expression of specific gene transcripts has been used successfully to provide environmentally relevant information. Advances in the handling and detection of nucleic acids have made a number of methods available for detection and quantification of gene transcription. In all protocols, the first step is to isolate total RNA using one of a range of commercial kits and reagents. Messenger RNA (mRNA) can then be separated using membrane or magnetic bead technology. This can then either be directly probed or used for RT-PCR after conversion to complementary DNA (cDNA) by reverse transcription. During all RNA handling steps great care must be taken to protect sample from degradation by ribonucleases (widespread RNA degrading enzymes). Typically RNA samples should be handled in a dedicated laboratory, using double autoclaved equipment and double autoclaved or certified RNAse free reagents. The range of potential detection methods for quantification of gene expression through mRNA measurement are detailed below.

3.3.2. Northern, blotting

Northern, dot and slot blotting are all hybridisation techniques. For Northern blotting, total RNA is electrophoretically resolved under denaturing condition and then transferred via capillary action to a nitrocellulose membrane. The membrane is probed with a radiolabelled oligonucleotide probe designed with a sequence that matches the target gene product. When a gene is highly expressed, more mRNA is present on the nitrocellulose and as a result there is greater hybridisation of the labelled probe. When viewed by exposure of audioradigraphic film to the membrane, such samples show a larger 'blot' than low expression samples. Image analysis can be used for formal quantification.

3.3.3. Dot and slot blotting

In dot and slot blotting, samples of cloned DNA matching the gene of interest are denatured and identical amounts uniformly spotted onto a single nitrocellulose membrane. The filter is then hybridised with a radioactivity labelled probe, containing the corresponding sequence in unknown amounts. The extent of hybridisation is estimated semiquantitatively by visual comparison to similarly spotted radioactive standards.

3.3.4. RT-PCR for measurement of gene expression

The power and specificity (through the design of gene specific primers) of PCR makes the procedure ideal for detecting responses of specific gene. Initially an obstacle to the use of PCR for gene expression quantification was the nature of amplification. This is characterized by a logarithmic increase to a plateau because the plateau is derived from limitation not related to the quantity of template, this means that parallel reactions with vastly different template inoculations result in near identical final product levels. As a result, comparisons of product levels at amplification end would show no difference between samples containing different amount of starting material. To overcome the problems of PCR for gene quantification, techniques have been developed that take a snap shot of product levels during reaction. Initially these quantitative PCR (QPCR) methods use either amplification for a limited number of cycles followed by gel electrophoresis and image analysis or quantitative competitive PCR. Such methods have now largely been supplanted by fluorescence in situ monitoring using specifically designed instruments. These platforms monitor PCR progress using two main fluorescence detection systems.

The recent technologies pyrosequencing and whole genome sequencing are dealt separately in the ensuing chapters.

3.4. TERMINOLOGIES RELATED TO SOIL METAGENOMICS

3.4.1. Genomics

The genome describes the full set of genetic "instructions" retained by an individual organism. Alongside the sequences that encode the building blocks of our cells "proteins" there are also instructions that control the expression of each gene in response to environmental change as well as significant stretches of redundant information. A strict definition of genomics would describe studies relating to the structure of genomic DNA. Today, therefore, genomics is a broadly used term encompassing numerous scientific disciplines and technologies.

These disciplines include: genome sequencing; assigning function to identified genes; determining genome architecture; studying gene expression at the transcriptome level (transcriptomics); studying protein expression at the proteome level (proteomics); and investigating metabolite flux (metabolomics) (Figure 3.3). Due to the magnitude and complexity of '- omic' data, these disciplines are underpinned by information technology support through bioinformatics.

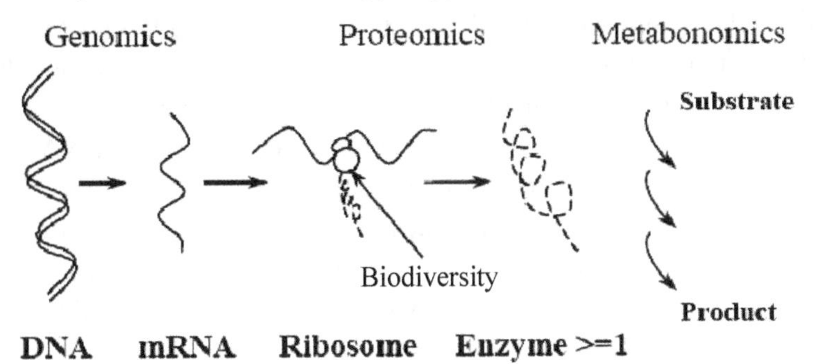

Fig. 3.3. The molecular basis of life

3.4.2. Transcriptomics

When a cell senses changes in its environment, it responds by accessing different components of its genome. For the most part, this access comprises the expression of genes encoding instructions for the production of new cellular "building blocks" or proteins. These instructions are read from the genome using a process called transcription to create messenger RNA (mRNA) molecules known as transcripts. Transcriptomics is loosely defined as the systematic quantification of the levels of all or a large proportion of the transcripts expressed within a cell under particular environmental conditions.

3.4.3. Microarray / macroarray

All "omics" approaches require the quantification of large numbers of biological parameters at one time. One method is to display the biological entities (probes) within a defined pattern or array. Arrays have been constructed displaying DNA, proteins and antibodies on a range of surfaces including membranes, glass slides and gels. Most commonly a reference to a microarray describes the precision dispensing of biological entities (probes) onto the surface of a charged membrane. Either through exposure to temperature or by UV cross-linking the probes are fixed within a defined location on the membrane. The biological entity is thus displayed within a defined grid from which a signal can be measured and attributed to each of the probes arrayed.

3.4.4. Proteomics

The building blocks of our cells are "proteins" which are encoded within our genomes and made by making an RNA copy of the genome sequence, this is then processed in Eukaryotes, to generate mRNA, which is in turn converted into protein by ribosomes in a process known as translation. However, proteins can be modified after synthesis or translation, in a process known as post translational modification, either through specific cleavage (proteolysis) or by the addition of biological molecules (sugars, phosphates etc). Therefore, although it is predicted that the human genome encodes only 40,000 transcripts, due to mRNA processing and posttranslational modification this may generate up to 10 times that number of discrete protein molecules. Proteomics is the global analysis of these cellular building blocks. Classically proteins have been separated using size fractionation through an acrilimide gel. More complex mixtures can be separated by exploiting the charge and size of a protein. By using this two dimensional approach, extremely complex mixtures of proteins can be separated. However, this separation only reveals the size and charge of the protein and does not facilitate its identification or allow it to be linked to the gene by which it is encoded. Recent advances in mass spectrometry have allowed extremely accurate size measurements to be performed on proteins separated using 2D electorphoresis.

3.4.5. Metabolomics / Metabonomics

Metabolites are classically defined as the small molecules or non-polymers that represent the substrates and products of the chemical reactions occurring within a cell. These small molecules range from well characterized compounds such as sugars, lipids, amino acids and nucleotides, to more novel structures. Metabolomics is the study of the levels and composition of these metabolites within a cell. There has been a divergence in the nomenclature and approach to this area of research. The terms metabolomics and metabonomics have both been used within this area. Although absolute definitions or distinctions are still being discussed, it is widely accepted that metabolomics describes the levels of native metabolites within a cellular system under varying environmental conditions, whilst metabonomics explicitly describes a systemic response profile taken from a sample collected from a complex, multi-cellular organism. In parallel with approaches in proteomics, many metabolomic techniques utilize methods that focus analysis within a discreet class of metabolites, therefore many researchers exploit a selective extraction procedure followed by gas or high pressure chromatography coupled to mass spectral analysis. These approaches provide profiles of sub or discreet metabolomes. In contrast, more global approaches have been exploited by using nuclear magnetic resonance to generate a fingerprint of the complete metabolomes. The advantage of the former is that it can be executed rapidly and with limited specialist apparatus whilst the facilities required for the latter are highly specialised and exist only within a handful of laboratories in the country. Furthermore, with appropriate standardization, the former can be used to identify specific metabolites relatively straightforwardly whilst the latter requires specialist knowledge and further analysis in order to allow metabolite identification.

3.4.6. Bioinformatics

When considering "omic" approaches, two distinct data issues become apparent. Firstly, a mathematical problem presents itself, connected with the statistical evaluation of the data and the interpretation and refinement of the results. Coupled with this is the association of each experiment with the relevant experimental description and related information concerned with each biological element being examined, this associated information is known as metadata. To address these issues it is essential to amalgamate mathematical expertise, computational competence and biological knowledge. Bioinformatics is the subject area that fills this essential interface. It is an essential component of any "omic" project and is the tool that will convert data overload into a useable output.

3.4.7. Toxicogenomics and eco-toxicogenomics

These are subdisciplines combining the fields of genomics and (mammalian) toxicology t has also been described as the study of genes and their products important in adaptive responses to chemical-derived exposures. The toxicogenomic approach presents important opportunities to improve understanding of the molecular mechanisms underlying toxic responses to environmental contaminants.

Mini Quiz

1. Mention the gold standard techniques used to examine microbial ecology *insitu*?
2. Differentiate between DGGE and TGGE?
3. What is FISH? How this technique is used in metagenomics and microbial ecology?
4. Elucidate the steps involved in single gene transcript quantification?
5. Differentiate between metabolomics and metabonomics?

CHAPTER - 4

ACCESSING THE SOIL METAGENOME

Direct cultivation or indirect molecular approaches can be used to explore and exploit the microbial diversity present in soil. Cultivation and isolation of microorganisms is the traditional method but, as only 0.1% to 1.0% of the soil bacteria are culturable using standard cultivation methods. Only a tiny portion of the gene pool has been characterized using cultivation and isolation and the majority remains unexplored

4.1. SOIL METAGENOME

Theoretically, the microbial DNA isolated from a soil sample represents the collective DNA of all the indigenous soil microorganisms, and is named the soil metagenome. The complex soil matrix, which contains many compounds (such as humic acids) binds to DNA and interfere with the enzymatic modification of DNA. So the recovery of microbial soil DNA that represents the resident microbial community and cloning or PCR is still an important challenge. In landmark studies, novel genes that encode useful enzymes and antibiotics were recovered by direct cloning of soil DNA into plasmid, cosmid or BAC (bacterial artificial chromosome) vectors and screening of the generated libraries. The genes were identified using functional screens and had little homology to known genes, which illustrates the enormous potential of soil-based metagenomic libraries. The same approach has been used to clone genes from soil communities that code for lipases, proteases, oxidoreductases, amylases, antibiotics, antibiotic resistance enzymes and membrane proteins.

The success to screen soil-derived metagenomic libraries depends on several factors: composition of the soil sample; collection and storage of the soil sample; the DNA extraction method used for high-quality DNA recovery; how representative the isolated DNA is of the microbial community present in the original sample; the host–vector systems used for cloning, maintenance and screening; and the screening strategy.

4.2. ISOLATION OF HIGH-QUALITY DNA FROM SOIL

4.2.1. Sampling of soil

Construction of a soil metagenomic library begins with sample collection. As soil samples are heterogeneous, details of physical, chemical and biotic factors such as particle size, soil type, water content, pH, temperature and plant cover are useful for evaluation and comparison of the outcomes of soil-based studies. Sampling is easier for surface soils compared with other environments such as subsurfaces. As microbial populations are large, sample volumes can be small (≤500 g in most studies). Disturbing soil during sampling might alter the composition of soil microbial communities, so the time that a sample is stored and transported should be kept to a minimum. A stored sample might not be representative of the undisturbed field soil.

4.2.2. Isolation of Metagenomic DNA

Library construction requires sufficient amounts of high-quality DNA which is representative of the microbial community present. Because of the heterogeneity of soils, the extent of microbial diversity and the adherence of microorganisms to soil particles, DNA extraction is particularly challenging. Methods described for metagenomic DNA isolation from soil and sediment samples can be broadly classified into direct and indirect extraction procedures. The ideal methodology should include unbiased lysis of cells and extraction of their nucleic acids.

Direct DNA isolation is based on cell lysis within the sample matrix and subsequent separation of DNA from the matrix and cell debris (Ogram *et al.*, 1987). The indirect approach involves the separation of cells from the soil matrix followed by cell lysis and DNA extraction (Holben *et al.*, 1988).The amounts of DNA isolated from different soil types using a selection of protocols range from less than 1 μg to approximately 500 μg of DNA per gram of soil. More DNA is recovered using the direct lysis approaches.

4.2.2.1. *Direct DNA isolation*

To achieve direct cell lysis, combinations of enzymatic treatment, high temperatures and detergent treatments have been used. In addition, several methods use mechanical disruption steps such as bead-beating, freeze–thawing or grinding of samples to lyse cells. In addition to the DNA that is recovered from lysed prokaryotes, extracellular DNA and eukaryotic DNA are also recovered. An excellent starting point is the direct lysis method which allows simultaneous recovery of DNA and RNA from soils of different composition.

i) **Cell lysis:** Cell lysis is a critical step in soil metagenomic DNA extraction. It is designed to release the DNA by breaking the cell wall and membranes of the microorganisms. Although the cell lysis efficiencies have been improved sufficiently by several investigators, complete cell lysis during DNA extraction still remains elusive and extraction bias still exists. Chemical or enzymatic lysis is relatively gentle and they often discriminate against particular cell types and do not completely penetrate soil or sediment samples. Mechanical disruption gives more uniform cell disruption and disperses soil or sediment samples to allow good penetration of the lysis buffer.Therefore; mechanical treatment is more effective and less selective than chemical lysis.

   The mechanical disruption methods include :

   a) Thermal shocks
   b) Bead-mill homogenization
   c) Microwave heating
   d) Ultrasonication.

   a) Thermal shock: consists of repeated freezing and thawing the sample suspensions. The number of freeze–thawcycles and the incubation time and temperatures can be varied. Thermal shock is less violent than other mechanical treatments such as microwave heating, ultrasonication and bead-beating.

   b) Bead beating: Extended beadbeater treatment results in progressively more shearing of DNA though the DNA yield increases proportionately. DNA yield increase with longer beating times,higher speeds and reduced extraction buffer volume. However, the DNA shearing always proportionately increased with DNA yield.

   c) Microwave heating: It is more efficient than the enzymatic lysis for the lysis of Gram-positive cells and spores. The succession of ultrasonication followed by microwave heating and thermal shocks was reported to be essential to achieve complete lysis of the *Streptomyces* spores.

d) *Ultrasonication:* This treatment efficiently releases the bacterial cells bound to the soil aggregates. Power and duration of sonication treatment are optimized based on the lysis efficiency and the shearing of DNA. The major disadvantage of severe mechanical treatments such as sonication and bead-beating is the shearing of DNA

ii) **Efficiency of cell lysis method:** Efficiency of a cell lysis method can be estimated by direct microscopic counts of soil smears obtained before and after the DNA extraction treatment. So far. lysis efficiencies upto 90 %by combining various methods has been reported where as grinding and beadbeating yields 25 to 66% and 74% respectively. The efficiency varies with the methods and the properties of soil. The lysis efficiency was negatively correlated with the soil clay content. Clay colloids in soils bind DNA and interfere with the recovery by trapping DNA in pelleted soil particles.

4.2.2.2. *Indirect DNA isolation*

Indirect DNA extraction methods based on cell separation, although less efficient in terms of the amount of DNA recovered, are less harsh than direct lysis methods. The separation of microorganisms from the soil matrix is achieved by mild mechanical forces or chemical procedures such as blending, rotating pestle homogenization or the addition of cation-exchange resins, followed by density gradient or differential centrifugation. Cells are then lysed with lysozyme and ionic detergent, and the DNA is extracted. The DNA obtained is almost entirely prokaryotic and seems to be less contaminated with matrix compounds, including humic substances. It minimizes the extraction of extracellular DNA, and provides larger fragment DNA with a high degree of purity. In addition, the average size of the isolated DNA is larger than that typically obtained by the direct lysis approach and is therefore more suitable for the generation of large-insert libraries. However, it has been reported that the DNA obtained only corresponds to about 25 to 35% of the total number of bacteria present in the soil. Different bacterial groups strongly adhere to soil particles, which might bias the picture of the composition of the microbial community in the sample.

Since no single method of cell lysis is appropriate for all soils, different combinations and modifications of lysis protocols may be needed for different soil samples. For example, sodium dodecyl sulphate (SDS) has been the most widely used detergent for cell lysis . However, the SDS-based lysis may not lyse some Gram-positive bacteria. The DNA yield was reported to be two- to six-folds higher for most of the Gram-positive bacteria by the grinding, freezing-thawing followed by SDS-based lysis. Thus, for soils exhibiting poor cell lysis or studies depending on extensive sampling of Gram-positive bacterial DNA, combinations of treatments should be considered. According to Gray and Herwig (1996) homogenization in a bead-beater for 1 min followed by incubation at 70 °C for 1 h in high salt-SDS buffer produced twice as much DNA as either the bead-beating or lysis at 70 °C alone.

4.3. METAGENOMIC DNA EXTRACTION

After cell lysis, many authors have used classical deproteinisation in organic solvents, *i.e.,* phenol, phenol–chloroform, chloroform–isoamyl alcohol before precipitating the metagenomic DNA. Proteins can also salted-out using saturated salt solutions such as sodium chloride, ammonium acetate, potassium acetate and sodium acetate. The proteins precipitate during the centrifugation at low speed and the nucleic acids are recovered in the supernatant. De-proteinisation with NaCl allows soil particles to precipitate with the cell debris and proteins. The pH of the extraction buffer also plays a vital role in the recovery of soil metagenomic DNA. However, larger amounts of humic material were released at pH 10.0 than at pH 9.0 and therefore, pH 9.0 has been proposed as the optimum pH of the extraction buffer.

DNA precipitation, which is performed to discard the extraction buffer and contaminants, is also a crucial step influencing the quality of metagenomic DNA. After deproteinisation, DNA precipitation of DNA is achieved with either isopropanol or ethanol. Polyethylene glycol (PEG) has also been widely used for the precipitation of soil metagenomic DNA. Alcoholic precipitation favors the coprecipitation of humic acids, while PEG greatly reduces the humic substance co-precipitation. However, PEG 8000 should be removed by phenol extraction as it might interfere with PCR. Alternatively, precipitation using 5% PEG yielded significantly less humic acids without affecting PCR and hence 5% PEG has been recommended for the precipitation of soil metagenomic DNA.

4.4. QUANTIFICATION OF METAGENOMIC DNA

Generally, A_{260} value is used to determine the concentration of DNA and the $A_{260/280}$ ratio is calculated to assess purity. $A_{260/280}$ ratio below 1.7 indicates the protein contamination. If the ratio is more than 1.7, DNA can be quantified based on an A_{260} value of 1.0 equivalent to 50 µg/ml (Sambrook *et al.*,1989). However, the humic substances interfere the quantification of soil DNA. It has been reported in many studies that A_{260} indicates the levels of humic substances rather than the DNA (Jackson *et al.*, 1997). Estimation of DNA concentration by the densitometric analysis of ethidium bromide (EtBr) stained agarose gel is another major approach. Therefore, spectrophotometric quantification of soil DNA is challenging due to the presence of organic compounds, which interfere the traditional estimation.

Alternatively, fluorometric analysis using the fluorescent dye PicoGreen offers efficient quantification of soil DNA. PicoGreen binds specifically to double-stranded DNA and the DNA-PicoGreen complex is quantified in a fluorometer. Using this approach, DNA concentrations from 25 pg/ml to 1 μg/ml can be quantified. Humic acid concentrations higher than 100 ng/μl interfere with the PicoGreen fluorescence. However, at concentrations below 10 ng/μl, humic acids do not interfere DNA quantification. Therefore, DNA sample can be diluted sufficiently, since it can quantify as low as 25 pg/ml, to estimate the quantity without the interference by humic compounds.Using this method, it is possible to measure crude DNA extracted directly from soil without any purification steps.

4.5. CONTAMINANTS OF METAGENOMIC DNA

Co-purification of contaminants such as humic compounds is a major problem. These contaminants are not completely removed during classical DNA extraction protocols, such as detergent, phenol-chloroform and protease treatments. Humic substances are formed by the decomposition of plant, animal and microbial biomass. Humic substances are structurally complex, polyelectrolytic, yellow to dark brown in color with the molecular mass range of 0.1 to >300 kDa. Based on solubility in acids and alkalis, humic substances can be divided into three main fractions. They are:

i) humic acid, which is soluble in alkali and insoluble in acid

ii) fulvic acid, which is soluble in alkali and acid

iii) humin, which is insoluble in both alkali and acid

Because of their heterogeneous nature, there is no single structure to define humic substances. The humic substances have threedimensional structures with the ability to bind other compounds to their reactive functional groups and absorb water, ions and organic molecules. Thus, almost all natural organic compounds can become bound or absorbed to humic substances. In addition, humic acids have physicochemical properties similar to that of nucleic acids. Therefore, humic substances along with the adsorbed organic molecules are generally co-extracted with DNA. Humic acids affect almost all molecular biological methods such as hybridization, restriction digestions of DNA, PCR and bacterial transformation.

4.5.1. Interference of humic substances in Polymerase chain reaction

Humic compounds interfere with PCR by inhibiting the DNA and Taq polymerase interaction. The phenolic groups of humic compounds denature biological molecules by bonding to amides or oxidize to form a quinone, which covalently bonds to DNA or proteins. Phenolic compounds from the sample or carried over from organic DNA purification procedures can also inhibit PCR by binding to or denaturing the polymerase . Dilution of samples provides a rapid and straightforward way of permitting PCR amplification. Diluting the sample exploits the sensitivity of PCR by reducing the concentration of inhibitors relative to target DNA. However, in few cases, the dilution may not be sufficient to decrease the humic acids inhibition. However, the mechanism of inhibition by humic substances is not yet elucidated.

Successful PCR amplification is generally used as an indicator of soil DNA purity. DNA extracted from soils can be tested for their suitability for PCR by performing an inhibition experiment based on the addition of soil DNA to a known PCR reaction mix. For instance, a standard 50 μl PCR reaction, designed to amplify an insert of a cloned gene using the recombinant plasmid as template with M13 primers, may be performed in the presence of different quantities of soil DNA . Since the PCR protocol can be sufficiently optimized by addition of enhancers or by the use of high-activity Taq polymerases, a more suitable purity marker is the successful cloning of metagenomic DNA. The cloneability of metagenomic DNA can be studied by restriction digestion and ligation efficiencies.

4.5.2. Quantification of humic compounds

Humic compounds absorb at 230 nm,whereas DNA absorbs at 260 nm. Therefore, absorbance ratio at 260/230nm (DNA/humic acid) and 260/280 nm (DNA/protein) are commonly used to evaluate the purity of the soil metagenomic DNA. Estimation of humic acids is influenced by the concentration of nucleic acids and protein contaminants with the absorbance in UV range. The absorbance at 320 nm can be used to measure the level of humic acids, which is independent of DNA and protein content. A_{320} values were strongly correlated (R2=0.911) with the level of PCR contamination.

There are two methods of measuring humic acid levels, which are i) absorbance at 340 nm and ii) fluorescence excitation at 471 nm and emission at 529 nm). Both absorbance reading at 340 nm and humic acid fluorescence yielded accurate and reproducible results over the humic acid concentration range of 0.1 to 100 ng/μl.

4.6. PURIFICATION OF METAGENOMIC DNA

Several strategies have been developed for the purification of soil metagenomic DNA. Caesium chloride density gradient centrifugation is a widely used and an efficient strategy for the purification of DNA from contaminants. But this method is not suitable for purification of multiple samples due to the longer processing time.Recently, several simple and rapid purification methods have been reported for the successful removal of contaminants from metagenomic DNA. These methods include the

a) Preprocessing of soil samples,

b) Agarose gel purification

c) Electroelution

d) Various chromatographical separations

a) **Removal of inhibitors by sample preprocessing:** Humic compounds compete with nucleic acids during DNA precipitation or adsorption site during the purification step using minicolumns. The addition of exadecyltrimethylammonium bromide (CTAB) or polyvinylpolypyrrolidone (PVPP) to soil-buffer slurry before cell lysis inhibited the co-precipitation of humic substances and improved the purity of metagenomic DNA. In microwave-based method also, addition of PVPP at high concentration in the extraction buffer has improved the purity of the extracted DNA by minimizing co-purification of humic substances. This method was successfully applied to different environmental samples such as activated sludge soil and sediments. Similarly, additions of sodium ascorbate or skim milk also decrease the co-extraction of humic acids. However, these compounds could not completely remove humic substances, since the reported A260/230 and A260/280 ratios for soil metagenomic DNA were significantly lower than the ratios for DNA isolated from pure cultures Therefore, additional purification steps may be required for the complete removal of contaminants.

   i) **Chemical flocculation:** The use of multivalent cations is a standard method for removing suspended organic solids during the purification of drinking water, a process referred as chemical flocculation. Hence, chemical flocculation with $AlNH_4(SO_4)_2$ can be considered as an alternative method for the elimination of PCR inhibitors. Though the mechanism of removal of organic inhibitors by $AlNH_4(SO_4)_2$ treatment is not known, it has been proposed that the aluminum cations may preferentially interact with the open and random structure of humic substances, while leaving the helical structure of DNA. Pretreatment of samples with calcium carbonate ($CaCO_3$) yielded PCR-compatible soil metagenomic DNA.

ii) **Treatment with powdered activated charcoal:** DNA extraction from polluted soil and sediments is more challenging due to the presence of metal ions and other pollutants in addition to humic substances. PAC has a vast application in the removal of various contaminants due to its extraordinarily large surface area and pore volume that gives it a unique adsorption capacity. The method involves gentle mechanical lysis, treatment with powdered activated charcoal (PAC) and ion-exchange chromatography with amberlite resin. Humic acids have a metal binding capacity and have been reported to form complexes with Pb and Cu Similarly, amberlite ion-exchangers are known to absorb the metal cations like Zn^{2+} and Cd^{2+}. Therefore, quality of the extracted DNA after this treatment is suitable for PCR and subsequent molecular diversity analysis

b) **Agarose gel purification:** Extraction of DNA from agarose gels after electrophoresis is a routine procedure in molecular biological experiment. Under standard electrophoretic conditions, humic acids comigrate with nucleic acids. The addition of PVPP to the agarose gel eliminates co-migration by retarding the electrophoretic mobility of humic compounds.

Similarly, a DNA extraction method, which was originally developed for pulsed-field gel electrophoresis (PFGE) has been proposed as a purification strategy. Basically, whole cells are embedded in low melting point (LMP) agarose blocks and then immersed in a lysis buffer. Severa lwashing steps can be carried out "in gel" prior to other treatments such as restriction digestion. High molecular weight DNA remains trapped within the agarose blocks, whereas cell debris and contaminants are freely diffused during lysis and washing steps. By this procedure, intact soil metagenomic DNA can be obtained without shearing and free of contaminants. In addition, agarose-embedded DNA can be directly used as template for PCR, since reactions are not affected by the presence of high quality LMP agarose concentrations even as high as 0.3% in the PCR mixture

c) **Electroelution:** Although agarose gel electrophoresis and PFGE can be used to purify soil DNA, EtBr staining, nucleic acid exposures to UV light and substantial sample processing time are considered to be disadvantageous. Therefore, electroelution has been considered as an alternate strategy for the soil DNA purification.

The standard electro elution involves elution of DNA fragments out of an agarose gel slice into a dialysis bag and subsequent precipitation. In general, a 2% or 4% LMP agarose gel solution is mixed with an equal volume of crude DNA and agarose blocks are electrophoresed. After the migration of humic compounds out of the agarose block, pure high molecular weight DNA can be electroeluted. Staining with EtBr is not required in this strategy.

The electroelution can also be used for the purification of RNA from humic substances. However, due to the smaller size of RNAs, electroelution protocol should be modified to avoid co-elution of humic contaminants. Addition of PVPP in the agarose blocks halt the migration of humics and thus pure RNA could be electroeluted first. Electroelution increases, the PCR sensitivity up to a factor of 10^4 relative to DNA templates before electroelution. Likewise, electroelution increased reverse transcription-PCR sensitivity by a factor 10^3 relative to unpurified RNA templates.

d) **Chromatographic techniques:** Various chromatographic techniques are employed for the purification of soil metagenomic DNA which are based on the principle of gel filtration, ion exchange and adsorption (Table 4.1).

Gel filtration also known as size exclusion resins are widely used for the purification of soil DNA. Typically, inhibitory substances are removed using spin columns packed with various gel filtration resins. Among the three different gel filtration resins (Sepharose 4B, Sephadex G-200, and Sephadex G-50) for the removal of humic contaminants, Sepharose 2B is superior in separation of DNA from humics and showed greater amplification than DNA purified by other materials. The crude DNA extracts could be easily purified using the Sepharose 2B column regardless of the lysis and extraction methods used. Using Sepharose 2B, greater than 90% of the DNA from a crude extract could be recovered while limiting humic acid recovery to less than 0.04%. The other methods along with the matrix and principle are given in Table 4.1.

Table 4.1. Various chromatographic methods for the purification of soil metagenomic DNA

Principle	Matrix	References
Gel filtration	Polyacrylamide Bio-gel P-6 and P-30 Sephadex 100, Sephadex G-75 Sephacryl S200 and S400, Sephadex G200 and G150, Sepharose 6B,4B and 2B Bio-Gel P100, P200 Toyopearl HW 55, HW 65, and HW 75	Tsai and Olson, 1992 Leff *et al.*, 1995 Cullen and Hirsch, 1998 Frostegard *et al.*, 1999 Miller, 2001
Ion exchange	DEAE Cellulose Amberlite IRA-400	Schneegurt *et al.*, 2003 Desai and Madamwar, 2007
Adsorption	Hydroxyapatite	Purdy *et al.*, 1996
	PVPP	Cullen and Hirsch, 1998
	Silica	Kauffmann *et al.*, 2004
	Powdered activated carbon (PAC)	Desai and Madamwar, 2007
Combined	Sephadex G200 and PVPP	Kuske *et al.*, 1998
	Sepharose 4B and PVPP	Arbeli and Fuentes,2007

4.7. PURIFICATION BY COMBINATION OF METHODS

Comparative evaluation of various methods by various investigators indicated that only a combination of two or more methods render sufficiently pure DNA from soil. Thus, removal of humic contaminants was most efficient with the combined column, probably due to two distinct mechanisms: size exclusion by Sepharose 4B and adsorption by PVPP. The consecutive filtrations with a Sepharose column followed by a PVPP column was superior to consecutive filtrations with Sepharose followed by Sepharose or PVPP followed by PVPP. In addition, Sepharose column followed by a PVPP column yielded better results than by the opposite order. Thus, the optimized purification protocol confirmed that only a combination of methods would be efficient in yielding pure environmental DNA from humic-rich soils.

Mini Quiz

1. Explain direct DNA isolation methods?
2. How can you overcome the interference of humic substances in PCR?
3. What is the principle involved in chemical flocculation?
4. Brief the methods involved in purification of metagenomics DNA?
5. What does the absorbance at 320 nm signifies?

CHAPTER - 5

CONSTRUCTION OF SOIL METAGENOMIC LIBRARIES

A variety of approaches may be employed for analyzing the soil metagenome, depending on the specific aims of the study. The ultimate downstream application should dictate the methods used for soil sampling, DNA extraction and purification, and library construction and screening (if necessary). The biologically, chemically, and physically heterogeneous nature of soils presents many challenges to the successful characterization of its microbial metagenome. Representative coverage of the soil microbial community requires isolation and cloning of a large amount of DNA from a small sample, and depends on insert size and the number of clones.

Constructing soil-based libraries involves the same methods as the cloning of genomic DNA of individual microorganisms; that is, fragmentation of the soil DNA by restriction-enzyme digestion or mechanical shearing, insertion of DNA fragments into an appropriate vector system, and transformation of the recombinant vectors into a suitable host. Although the generation of soil libraries is conceptually simple, the size of the soil metagenome and the large number of clones that are required for full coverage make this a daunting task.

The major breakthrough in soil metagenomics was the construction of libraries from soil DNA and screening of these libraries by functional and sequence-based approaches. This technology paved the way for elucidating the functions of organisms in soil communities, for genomic analyses of uncultured soil microorganisms and for the recovery of entirely novel natural products from soil microbial communities. The general protocols for the metagenomic DNA isolation and purification should be suitably modified based on the objectives. The goals of metagenomic projects may be the i) identification of functional genes, ii) estimation of the microbial diversity, iii) understanding the population dynamics of a whole community or iv) assembly of complete genome of an uncultured organisms. For example, when bead mill homogenization is selected, parameters should be set based on the intended use of metagenomic DNA. If the DNA is to be used to construct gene banks, large fragments are required in order to minimize the number of clones that need to be screened. On the other hand, if the DNA is to be used in PCR, DNA fragment size may not be as important as DNA yield. The major factors to be considered while isolating soil metagenomic DNA for various applications are discussed in this chapter.

5.1. LIBRARY BIAS AND DNA EXTRACTION

As different soil microorganisms have different susceptibilities to cell lysis methods, the sequences present in the isolated DNA and the libraries is dependent on the extraction method. How much bias in libraries is due to extraction methods has not been studied intensively. It is usually presumed that the DNA isolated by the direct lysis approach better represents the microbial diversity of a soil sample because this method does not include a cell separation step, so microorganisms that adhere to particles are also lysed. Direct lysis approaches have been used more frequently than the separation techniques to isolate soil DNA for the construction of libraries.

5.2. LIBRARY SIZE

Libraries can be classified into two groups with respect to average insert size: small-insert libraries in plasmid vectors (less than 15 kb) and large-insert libraries in cosmid, fosmid (both up to 40 kb) or BAC vectors (more than 40 kb). The host for the initial construction and maintenance of almost all published libraries is *Escherichia coli*. Shuttle cosmid or BAC vectors can be used to transfer libraries that are produced in *E. coli* to other hosts such as *Streptomyces* or *Pseudomonas* species. The choice of a vector system depends on the quality of the isolated soil DNA, the desired average insert size of the library, the vector copy number required, the host and the screening strategy that will be used.

Soil DNA that is contaminated with humic or matrix substances after purification or DNA sheared during purification might only be suitable for production of plasmid libraries. Small-insert soil-based libraries are useful for the isolation of single genes or small operons encoding new metabolic functions. Large-insert libraries are more appropriate to recover complex pathways that are encoded by large gene clusters or large DNA fragments for the characterization of genomes of uncultured soil microorganisms. It has been estimated that more than 10^7 plasmid clones (5 kb inserts) or 10^6 BAC clones (100 kb inserts) are required to represent the genomes of all the different prokaryotic species present in one gram of soil. These estimates are based on the assumption that all species are equally abundant. To achieve substantial representation of the genomes from rare members (less than 1%) of the soil community, it has been calculated that libraries containing 10,000 Gb of soil DNA (10^{11} BAC clones) is needed in library construction. It is not reasonable to suppose that bacterial taxa present in lower abundance will be represented within a metagenomic library unless an enrichment method is used. In addition, a comparison of the 16S rRNA genes in a BAC library with a collection of DNA fragments that were generated by direct PCR amplification and cloning of the 16S rRNA genes from the same soil sample indicates that the representation of certain bacterial groups in the library iss different from that present in the soil sample. Also, when working with soil samples that have not been well-characterized, it is advisable to utilize a variety of different methods for DNA extraction and purification to empirically determine the ideal combination that will yield high-quality and high-diversity metagenomic DNA.

Here, we discuss many of the metagenomic-based approaches used to study soil microbiology. Two approaches, the function-driven analysis and the sequence-driven analysis, have emerged to extract biological information from metagenomic libraries

5.3. LIBRARY CONSTRUCTION

a) **Expression libraries:** Metagenomic DNA constitutes a promising source of novel metabolites. Potential discovery of these compounds requires the construction of metagenomic DNA library and its efficient screening. Three parameters have to be carefully considered when constructing metagenomic libraries.

   i) The large size of the metagenome, is defined as the combination of all genomes from bacteria colonizing a given environment. According to reassociation kinetic data, the genetic diversity of the soil metagenome is between 5000 and 5,000,000 fold higher than that of the *E. coli* genome. Such diversity requires improved cloning efficiency so that the clones in the gene library provide an acceptable representation of the entire metagenome. Thus, in view of the large number of prokaryotic species present in soil, large-scale cloning techniques need to be used to cover the collective genomes, which require substantial amounts of high-quality metagenomic DNA.

ii) The second consideration is the size and cluster organization of genes involved in the synthesis of secondary metabolites. Since the genes encoding the biosynthesis of secondary metabolites are frequently clustered, metagenomic libraries should contain clones with larger DNA inserts. Isolation of high molecular weight DNA facilitates the cloning of DNA into bacterial artificial chromosomes (BACs) and allows the characterization of large regions of the genomes. In general, the indirect methods yield higher molecular mass DNA with greater purity than direct methods. Direct DNA isolation methods from soil and sediments are hampered by the problems of mechanical shearing due to physical forces imposed on the sample during isolation such as bead beating. Furthermore, nucleases released during cell lysis may degrade the released DNA.

iii) Eukaryotic DNA content is also a critical parameter for the construction of expression gene banks in *E. coli* or other bacterial hosts. Since the expression of eukaryotic genes in bacterial hosts is limited, eukaryotic DNA will increase the number of clones that need to be prepared and screened. This negative effect is still increased by the generally much larger genome size of eukaryotes. Depending on the soil sample used, direct DNA extracts contained 61–93% of eukaryotic nucleic acids, which may be due to the partial lysis of indigenous eukaryotic organisms such as fungi, algae, and protozoa, or it may be caused by lysis of residual plant material. In contrast, DNA obtained by indirect method was primarily derived from bacterial cells (92%) due to the separation from eukarya by differential centrifugation, which makes it suitable for expression cloning. Therefore, it is emphasized that direct lysis should be avoided when gene banks are constructed in bacterial hosts.

Despite these limitations, analysing and screening of libraries has yielded several novel biomolecules and provided insights into the genomes of uncultured prokaryotic soil organisms and the ecology of the soil ecosystem.

b) **Sub-metagenomic libraries:** Construction of selective sub-metagenomic libraries will be useful for the easier identification of genes from different groups of microorganisms incase of function based screening. It is known that *E. coli* does not recognize approximately 80% of actinomycete promoters. Due to the difference in G+C contents also, most of the actinomycete genes are unlikely to be expressed in *E. coli*. However, for functional screening, the number of functionally expressed genes should be as high as possible.

Therefore, DNA from organisms that can barely be expressed may be excluded in the respective host and thus, it is desirable to exclude actinomycete DNA when *E. coli* is used as host. Nevertheless, actinomycetes possess interesting genes coding for enzymes and antibiotic synthesis.

Actinomycetes devote a large part of their genomes to the synthesis of secondary metabolites. An average actinomycetes strain has the genetic potential to produce 10–20 secondary metabolites. Thus, it is important to develop metagenomic libraries for the identification of functional genes from environmental samples that are rich in actinomycetes. For this, a submetagenomic library may be created using *Streptomyces* as host. To make expression libraries in *Streptomyces*, the *E. coli–Streptomyces* artificial chromosome (ESAC) vectors have been developed.

Thus, metagenomic DNA library construction in *E. coli* after the selective elimination of actinomycete DNA with high G+C content increases the number of positive colonies. Similarly, DNA from low G+C bacteria can be selectively eliminated if a library is to be constructed in *Streptomyces*. The extracted total metagenomic DNA can be subjected to ultracentrifugation and separated based on G+C content. Though it may not provide a complete separation, it will certainly increase the representation of concerned group of genomes in the sublibrary. Alternatively, lysis methods can be optimized for the selective lysis of eubacteria by enzymatic methods followed by the lysis of actinomycete by severe mechanical treatments. In few cases, a particular metabolic process of an uncultured organism is of interest. Stable-isotope probing (SIP) holds a strong potential to selectively isolate DNA from organisms that can metabolize a particular substrate (See SIP in Chapter 6).

5.4. LIBRARIES BASED ON INSERT SIZE

5.4.1. Small-insert libraries

The construction and analysis of small-insert metagenomic libraries (less than ~10 kb average insert size) is a useful approach to identify gene product(s) encoded by a relatively small genetic locus, such as most enzymes, or genetic determinants of antibiotic resistance. Biases in cell lysis and cloning techniques may select against some prokaryotic taxa or gene products that are toxic to the host cell; therefore, it is important to select DNA extraction and cloning methods designed to yield a high proportion of DNA from the microorganisms of interest.

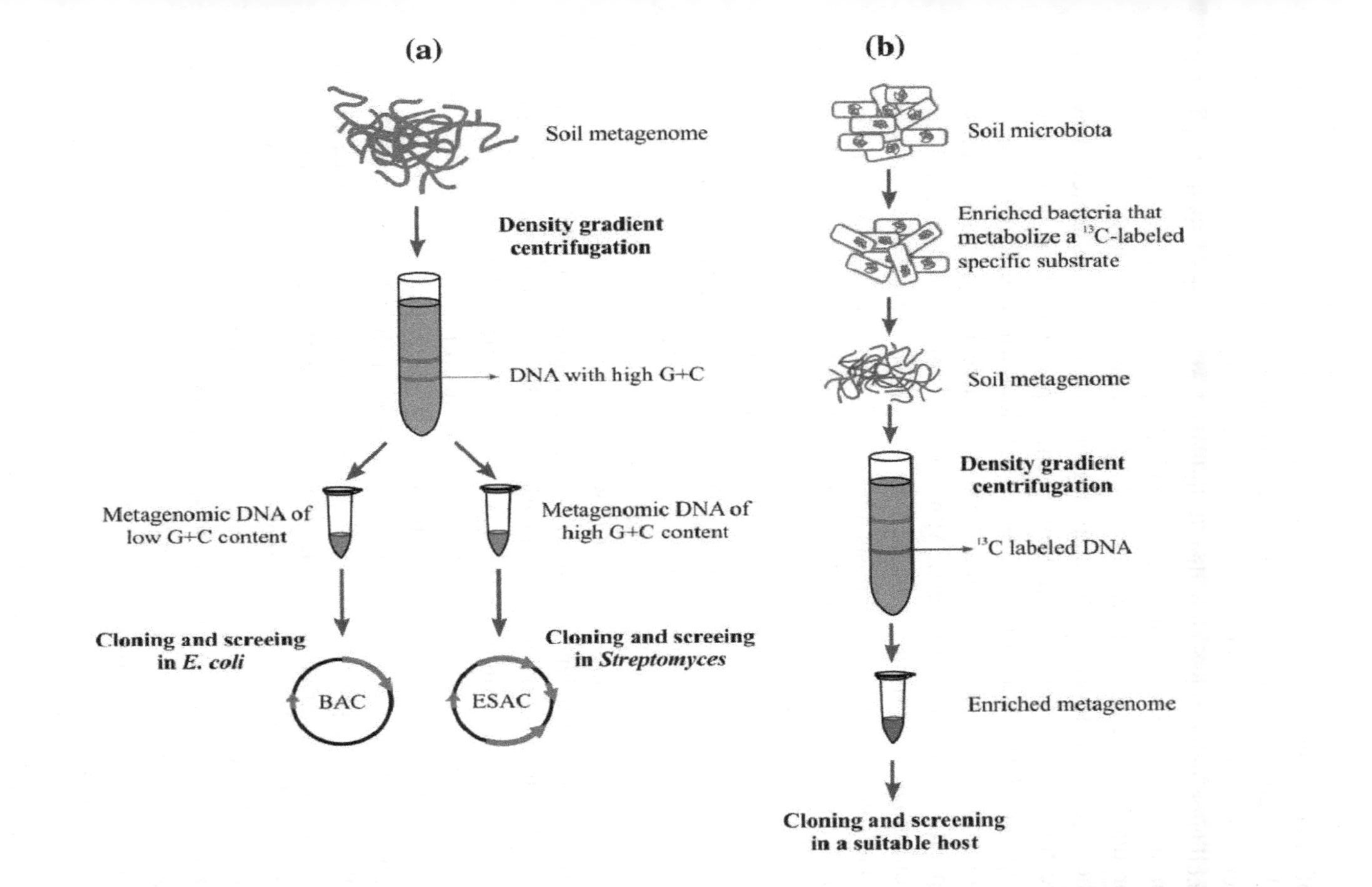

Fig. 5.1. Construction of sub-metagenomic libraries. (a) Separation of metagenomic DNA by G+C content; (b) Selective enrichment of DNA from bacteria that can metabolize a specific substrate by stable-isotope probing (SIP).

i) *Selection of vectors and host organisms*: Vectors used for the construction of small-insert libraries often possess a promoter for transcription of the cloned gene inserts and should be compatible with the host selected for screening. A vector with two promoter sites flanking the multi-cloning site facilitates gene expression that is independent of gene orientation and the promoters associated with inserts. With the possibility of the expressed gene product having toxic effects on the host organism, it is important to regulate the expression levels of the cloned genes, which can be achieved by using vectors with inducible control over gene expression of the insert or plasmid copy number. An additional issue to consider when selecting a vector is its ability to replicate in multiple hosts to enable heterologous expression of specific gene(s) of interest. Although the utility of using *E. coli* as a heterologous host for metagenomic library construction has been well established, other bacterial hosts may be more suitable for some applications, particularly if the percent G+C content of the cloned gene(s) are significantly different from that of *E. coli*, or if the regulatory factors required for expression or the biosynthetic capacity may be enhanced within another prokaryote.

ii) *Preparation of DNA for cloning*: The preparation of DNA for small-insert libraries is similar to that used for PCR- or pyrosequencing-based applications. A sufficient yield of DNA is necessary for successful library construction, and soil contaminants co-isolated with the DNA such as humic acids can interfere with efficient cloning. DNA extraction and purification conditions should be harsh enough to lyse a variety of microbes and remove the majority of contaminants, while the degree of DNA fragmentation that is permissible will depend on the desired average insert size of the library. If the desired average insert size is less than 20 kb, a commercial kit (e.g., MoBio Laboratories, Qiagen) may provide a useful method for obtaining DNA of sufficient size, purity, and yield for small-insert cloning. In cases when commercial kits are not suitable, such as soils with high clay content, it may be advisable to adopt cell based ("indirect extraction") methods such as sucrose/Percoll density gradient centrifugation or Nycodenz treatment*, which have been shown to generate DNA appropriate for small-insert cloning. Regardless of which DNA extraction method is used, it is possible that further purification will be required for efficient cloning. Many DNA purification methods may be effective in yielding DNA suitable for cloning, such as phenol and chloroform extraction, and/or treatment with hexadecyltrimethylammonium bromide (CTAB) or polyvinylpolypyrrolidone (PVPP), when combined with CsCl density centrifugation or hydroxyapatite column chromatographic purification. These methods were very well discussed in Chapter 4.

In the case of indirect extraction methods, some studies have found that a washing step prior to cell lysis is useful for the removal of soluble inhibitors and extracellular DNA. Unfortunately, many soil samples require a combination of these purification steps, which significantly increases processing time and can lead to an even greater loss of DNA. Following extraction and purification of the DNA, it may be physically sheared or partially restriction digested and then size-selected by extracting the DNA in the desired size range from an agarose gel. Because the size selected DNA will likely be less than 20 kb, it can be column-purified, the gel slices may be treated with GELase enzyme, or the DNA may be electroeluted from the gel prior to cloning. *(*Nycodenz treatment:* Nycodenz is a water soluble non ionic and non toxic derivative of benzoic acid. It is used as a gradient medium to separate cells based on the basics of buoyancy).

5.4.2. Large-insert libraries

Large-insert metagenomic libraries contain large, contiguous DNA fragments that have the potential to contain intact biosynthetic pathways involved in the synthesis of antimicrobial compounds, multiple enzymes with catabolic activity, or operons encoding other complex metabolic functions. However, along with potential advantages for some applications, large insert cloning from soil microorganisms also presents many technical challenges in order to obtain and screen high-quality metagenomic libraries containing DNA from representative microorganisms.

i) *Selection of vectors and host organisms*: Because the applications appropriate for large-insert metagenomic libraries depend on their ability to capture large, intact genetic pathways, the selection of an appropriate cloning vector is critical to the maintenance and expression of the cloned pathways. Several vector options exist for cloning high molecular weight (HMW) DNA from environmental samples, such as cosmids, fosmids, and BACs.

   a) *Cosmid*: a hybrid plasmid that contains *cos* sequences from the λ phage genome; was one of the first vectors used for cloning. The packaging capacity of cosmids varies depending on the size of the vector itself but usually lies around 40-45 kb. While typical plasmids can maintain inserts of 1-20 kb, cosmids are capable of containing DNA inserts of about 30 kb up to 40 kb. The size limits ensure that vector self-ligation resulting in empty clones is not a problem. Both broad host range cosmids and shuttle cosmids are available. Cosmids can replicate like plasmids when they contain a suitable origin of replication and they commonly possess selective genes such as antibiotic resistance to facilitate screening of transfected cells.

b) *Fosmid vectors*: are similar to cosmids but are based on the *E. coli* F-factor replicon, were developed for constructing stable libraries from complex genomes. The low copy number of fosmid vectors offers higher stability than comparable high-copy number cosmids. A low copy number is optimal for long term survival of the plasmid in a host. Also, plasmid copy number determines gene dosage. Recombinant clones from large-insert libraries may express gene products that are toxic to the host and hence it is important to maintain libraries in single copy until screening for a function. Fosmid copy number is tightly regulated in *E. coli* to 1-2 copies per cell, and fosmids can typically accommodate cloned inserts between 40 and 50 kb.

c) *BAC vectors*: These are based on the same F-factor replicon but have the capacity to maintain large inserts in excess of 100 kb. Along with the long-term stability conferred by the F-factor for maintenance, a modified BAC vector also containing an RK2 origin of replication is capable of inducible copy-number, alternating between single-copy and high-copy BAC maintenance. The inducible-copy phenotype can have significant advantages for the yield of DNA from metagenomic clones, and potentially for expression of cloned genes.

Although fosmid vectors are limited in insert size compared to BAC vectors, their significantly higher cloning efficiency enables construction of metagenomic libraries with many thousands of transductants. Conversely, BAC vectors even though capable of accommodating higher insert sizes have lower cloning efficiency than that of fosmid vectors. As mentioned previously, HMW DNA for fosmid-based cloning may be treated with harsher extraction and purification methods, which could yield a higher concentration of DNA from more diverse microorganisms than that of DNA isolated for BAC-based cloning. However, because BAC vectors can stably maintain cloned inserts hundreds of kilobases in size, they offer a greater chance of isolating intact pathways or of linking phylogenetic and functional genetic information. Therefore, the predicted size of the pathway of interest, its native level of activity, and its relative abundance within the community must be considered when choosing a suitable cloning vector for large-insert metagenomic library construction.

As with small-insert libraries, *E. coli* is the preferred host for the construction of large-insert metagenomic libraries due to its high cloning efficiency. This host has been successfully used to express many bioactive enzymes and compounds in metagenomic studies. In addition, *Streptomyces lividans* has been used as a heterologous host for library screening, and it has more stringent promoter recognition and regulation properties when compared to *E. coli* Because large-insert libraries may contain clones that express gene products that are toxic to the library host, it is important to maintain libraries in single copy until screening for a function and to consider the use of multiple hosts to increase the probability of identifying and characterizing the function(s) of interest. It has been shown that clones positive for a specific activity detected using one host may not be detected in a different host and *vice-versa*. A range of Gram-positive and -negative bacteria can be used as hosts for heterologous expression, and the corresponding vectors selected should be compatible with those hosts. Vector systems such as pRS44 enable shuttling into other Gram-negative hosts and have higher potential for function-based screening across species barriers and heterologous gene expression. Several other factors are necessary for successful expression of the cloned pathways (e.g., co-factors, post-translation modification enzymes, inducers, chaperones etc.), which may be provided by the vector or the host organism.

ii) **Preparation of DNA for cloning:** Large-insert metagenomic libraries are the most challenging to construct, but also can provide significant advantages for some applications since they enable identification and characterization of intact functional pathways encoded on large, contiguous DNA fragments. All of the considerations discussed previously regarding the selection of DNA extraction and purification methods apply to large-insert cloning, along with an additional critical issue: the construction of large-insert metagenomic libraries depends on obtaining sufficiently pure DNA of high molecular weight (in excess of ~100 kb). However, most extraction and purification methods result in DNA significantly smaller than this size. Although a few methods can yield DNA from soil greater than 1 Mbp in size it has been demonstrated that these indirect extraction methods can result in inefficient cloning due to contaminants that may be co-isolated with the metagenomic DNA and require further purification.

The successful recovery of high molecular weight (HMW) metagenomic DNA from soil microorganisms presents many extraction and purification challenges. A primary goal is to obtain DNA from an assemblage of diverse bacterial cells that are representative of the soil microbial community DNA. However, the harsh extraction methods (i.e., bead-beat lysis) typically employed for PCR or small-insert cloning applications will result in substantially fragmented DNA that is much too small for large-insert cloning. The use of indirect DNA extraction methods can somewhat alleviate this dilemma by first separating the cells from the soil sample, embedding them in an agarose plug, and then carefully lysing the cells and purifying the resulting DNA rather than performing the extraction *in situ*

The choice of extraction and purification method also depends on which cloning vector will be employed, such as a fosmid or bacterial artificial chromosome (BAC). Metagenomic libraries constructed in a fosmid vector are introduced into their heterologous host using a λ phage-based packaging system, which limits the clone insert size to 40-50 kb. Although DNA isolated for fosmid libraries must be treated carefully to prevent excessive shearing of DNA, using a fosmid vector does allow the use of harsher extraction and purification methods than those that may be used for BAC cloning. Also, during fosmid library construction, the DNA is typically size-selected by physically shearing the DNA into fragments of a desired length rather than by restriction digestion. This "direct size-selection" method eliminates the need for gel extraction (which can lead to DNA loss) and the possibility of DNA degradation due to over-digestion. An alternative to the physical shearing method was proposed by Quaiser and colleagues, who constructed fosmid libraries containing soil metagenomic DNA contaminated with humic and fulvic acids by embedding the DNA in agarose, electrophoresing the DNA through agarose containing PVP, and then combining the subsequent removal of the PVP with the size-selection step which resulted in purified, "clonable" DNA in the 30-100 kb size range. In combination with other purification steps, the inclusion of a formamide plus NaCl treatment was shown to significantly increase the efficiency of cloning of large DNA fragments into fosmid or BAC vectors.

Factors that have been demonstrated to affect the size of recovered DNA include not only the DNA extraction method used but also the microbial growth status and chemical. In general, DNA extracted from bacterial cells is significantly larger than DNA directly extracted from soil but is also found in lower yields however, this loss can be reduced by using wide-bore pipette tips to prevent shearing of DNA, performing multiple rounds of indirect extraction on each soil sample, minimizing the amount of agarose that is retained during size selection, or using electroelution as an alternative to extraction of DNA from the agarose gel. The merits and demerits of constructing small insert and large insert libraries are given in Table 5.1.

Table 5.1. Pros and cons of small-insert and large-insert soil libraries

Advantages	Disadvantages
Small-insert library (plasmids)	
High copy number allows detection of weakly-expressed foreign genes	Small insert size
Expression of foreign genes from vector promoters is feasible	Large numbers of clones must be screened to obtain positives

[Table Contd.

Contd. Table]

Advantages	Disadvantages
Cloning of sheared DNA or soil DNA contaminated with matrix substances is possible	Not suitable for cloning of activities and pathways that are encoded by large gene clusters
Technically simple	
Large-insert library (cosmids, fosmids, BACs)	
Large insert size	Low copy-number might prevent detection of weakly-expressed foreign genes
Small numbers of clones can be screened to obtain positives	Limited expression of foreign genes by vector promoters
Suitable for cloning of enzyme activities and pathways that are encoded by large gene clusters	Requires high-molecular soil DNA of high purity for library construction
Suitable for partial genomic characterization of uncultured soil microorganisms	Technically difficult

Mini Quiz

1. What is the major criterion involved in selecting vectors for metagenomics library construction?
2. What is GC content? How it influences the number of positive colonies if you intend to construct library of actinomycetes?
3. Differentiate between fosmids and cosmids?
4. What is the significance of RK2 region in BAC vector?
5. What is “direct size selection”?

CHAPTER - 6

ENRICHMENT STRATEGIES OF SOIL METAGENOME

The major bottleneck with analysis of metagenomic libraries is the low frequency of clones of a desired nature. To increase the proportion of active clones in a library, several strategies have been designed to enrich for the sequences of interest before cloning. The potential power of this strategy is evident in the elegant genomics performed on uncultured Bacteria and Archaea that are highly enriched in associations with their hosts.

The first complete genome sequence obtained for an uncultured bacterium is *for Buchnera aphidicola*, an obligate symbiont of aphids. The prokaryotic cells were separated from insect tissue to produce relatively pure microbial DNA, making it feasible to sequence and reassemble the genome despite the inability to grow the bacterium in culture. Genomic analysis has also been successfully conducted on the uncultured Archaea *Cenarchaeum symbiosum*, which is a symbiont of a marine sponge. *C. symbiosum* is highly enriched in the sponge – it is the only archaeal phylotype found in the sponge and represents 65% of the prokaryotic cells. DNA prepared from archaeal cells were further enriched by differential centrifugation. The successful assembly of the *B. aphidicola* genome as well as the *Vibrio cholerae* genome, which contains two chromosomes provides a strong foundation for developing biological and computational methods for assembling more complex genome assemblages. However, the first and essential step is to enrich for the genomes of interest. The different enrichment strategies proposed to improve the odds while screening metagenomic libraries are.

i) Sample enrichment
ii) Nucleic acid extraction and enrichment technologies
iii) Genome and gene enrichment
iv) Enrichments for metagenomic clones forming metabolic consortia

6.1. SAMPLE ENRICHMENT

In a metagenomic screening process (e.g. expression screening of metagenomic libraries), the target gene(s) represent a small proportion of the total nucleic acid fraction. Pre-enrichment of the sample thus provides an attractive means of enhancing the screening hit rate. The discovery of target genes can be significantly improved by applying one of several enrichment options (discussed below), ranging from whole-cell enrichment, to the selection and enrichment of target genes and genomes. For example, in the Sargasso Sea genome sequencing project size-selective filtration effectively removed the eukaryotic cell population. Alternatively, differential centrifugation has been used to enrich for *Buchnera aphidicola* and *Cenarchaeum symbiosum* symbionts by removing them from their hosts in preparation for whole genome sequencing.

Culture enrichment on a selective medium favours the growth of target microorganisms. The inherent selection pressure can be based on nutritional, physical or chemical criteria, although substrate utilization is most commonly employed. For example, a four-fold enrichment of cellulose genes in a small insert expression library was obtained by culture enrichment on carboxymethyl cellulose. Although culture enrichment will inevitably result in the loss of a large proportion of the microbial diversity by selecting fast-growing culturable species, this can be partially minimized by reducing the selection pressure to a mild level after a short period of stringent treatment.

6.2. NUCLEIC ACID EXTRACTION AND ENRICHMENT TECHNOLOGIES

More recently numerous community nucleic acid extraction methods have been developed. The two principal strategies for the recovery of metagenomic DNA are cell recovery and direct lysis. Extraction of total metagenomic DNA is necessarily a compromise between the vigorous extraction required for the representation of all microbial genomes, and the minimization of DNA shearing and the co-extraction of inhibiting contaminants. Mechanical bead beating has been shown to recover more diversity compared with chemical treatment. However, chemical lysis is a more gentle method, recovering higher molecular weight DNA. Chemical lysis can also select for certain taxa by exploiting their unique biochemical characteristics. The technologies for recovering RNA from environmental samples are largely similar to those used for DNA isolation, modified to optimize the yield of intact mRNA by minimizing single-stranded polynucleotide degradation. Protocols are designed to limit physical degradation and RNase activity, which are the major causes of yield loss. Samples should be processed or frozen at -80°C immediately after harvesting and additional methods used to minimize RNA degradation, such as the co-precipitation of cellular RNA with proteins (e.g. sulphate salt solution RNAlater,) and the synthetic capping of the isolated RNA. mRNA recovery has been applied extensively to eukaryotes, but has only recently been used in the study of prokaryote metagenomes. These techniques provide a feasible route for the construction of metagenomic cDNA libraries for the further identification of functional eukaryotic genes.

Total DNA extracted directly from environmental samples does not typically contain an even representation of the population's genomes within the sample. Rare organisms will contribute a relatively low proportion of the total DNA and the genome population might be over shadowed by a limited number of dominant organisms. This could lead to a selective bias in downstream manipulations such as PCR. This problem can be partially resolved by means of experimental normalization. Separation of genotypes is achieved by caesium chloride gradient centrifugation in the presence of an intercalating agent, such as bis-benzimide, for the buoyant density separation of genomes based on their % G and C content. Equal amounts of each band on the gradient are combined to represent a normalised metagenome.

Normalisation can also be achieved by denaturing fragmented genomic DNA, and re-annealing under stringent conditions (e.g. 68°C for 12–36 h). Abundant ssDNA will anneal more rapidly to generate double-stranded nucleic acids than rare DNA species. Single-stranded sequences are then separated from the double-stranded nucleic acids, resulting in an enrichment of rarer sequences within the environmental sample.

6.3. GENOME AND GENE ENRICHMENT

Genome enrichment strategies can be used to target the active components of microbial populations. Soil is a complex body containing an estimated 10^9 prokaryotes and more than 2000 genome types per gram of soil, with an average genome type representing less than 0.05% of the soil community. With today's technology it would be difficult to obtain complete coverage of all of the genomes in a soil community. Therefore, carving out slices of the community, selected for a common feature, reduces the complexity of the task and brings into sight the possibility of complete coverage of a subset of the soil community. A simple enrichment is for GC content of the genomes. As many organisms that have a high GC content in their DNA are of particular interest (e.g. *Actinomycetes* and *Acidobacteria*). DNA can be extracted from the soil and then subjected to ultracentrifugation to enrich for high GC content DNA. Although this is a fairly crude approach and will not provide a complete separation, it will certainly increase the representation of certain genomes in a library. The techniques involved are:

1) Stable-isotope probing (SIP)
2) Suppressive subtraction hybridisation (SSH)
3) Differential Expression analysis (DEA)
4) Affinity capture
5) Microarray
6) Phage display

6.3.1. Stable-isotope probing (SIP)

This technique involves the use of a stable isotope-labelled substrate and density gradient centrifugal separation of the 'heavier' DNA or RNA. Stable-isotope probing (SIP) s provides a ^{13}C-labeled substrate to soil bacteria (see Fig 6.1b). The bacteria that can use the substrate incorporate the ^{13}C into their DNA, making it denser than normal DNA containing ^{12}C. SIP has been successfully used for labeling and separating DNA and RNA. Density gradient centrifugation cleanly separates the labeled from unlabelled nucleic acids, which can then be used either for PCR-based analysis or direct cloning to construct metagenomic libraries. The method has enormous potential for subdividing microbial communities into functional units to simplify analysis and will offer broad opportunity to study community functions if it can be expanded to stable isotopes of other elements, such as phosphorus or nitrogen.

$^{13}CH_3OH$-labelling of forest soil metagenomic DNA resulted in the identification of both known α-proteobacterial methylotrophs and novel methanol dehydrogenase (*mxa*F) gene variants belonging to Acidobacterial taxa. Analysis of ^{13}C-phenol enriched anaerobic bioreactor populations by RNA-SIP demonstrated that phenol degradation was dominated by a member of the genus Thauera, a group previously unknown as phenol degraders. Other important technique associated with SIP is bromodeoxyuridine (BrdU) enrichment.

6.3.1.1. *Bromodeoxyuridine (BrdU) enrichment*

A more elegant separation method is bromodeoxyuridine (BrdU) enrichment (Figure 6.1.a). The principle underlying this type of enrichment is that metabolically active organisms will incorporate a labeled nucleotide into their DNA or RNA, which can then be separated by immunocapture or density gradient centrifugation. Addition of substrates with BrdU selects among the members of the microbial community for enhanced growth on the specific substrate.16S rRNA analysis performed on the DNA identifies phylogenetic groups that were metabolically active in the original sample. The addition of selective substrates with the BrdU further discriminates among the members of the microbial community, enriching metagenomic libraries for those that grow on the added nutrient.This strategy could be used to enrich for organisms that grows on xenobiotics or on substrates such as starch, cellulose and proteins to find amylases, cellulases and proteases, respectively, or other enzymes of interest in metagenomic libraries.

Many ^{13}C-, ^{18}O- and ^{15}N- labelled fine chemicals are available (e.g. phenol, methanol, ammonia, methane, carbonate etc.) but the wide application of SIP is limited by the commercial availability of complex labelled compounds that require expensive custom synthesis. BrdUTP labelling offers an alternative in cases where SIP labelled compounds are not available. Growth in the presence of BrdUTP and the unlabeled compound accesses metabolically active organisms.

6.3.1.2. *Limitations of these approaches*

Although these approaches provide a significant step in refining metagenomic library construction by enriching for DNA from a subset of the community, both methods have limitations that need to be addressed by future research.

a) Cross-feeding: Any bacterium that is metabolically active will take up BrdU and will therefore be represented in a library that is supposed to be enriched for organisms that utilize an added substrate.

b) Timing: The more prolonged the substrate feeding, the higher the probability that the substrate will be recycled in the community and the basis of the enrichment will break down.

c) The immunocapture method and density gradient centrifugation may shear the DNA, making it difficult to retrieve pathways encoded on large fragments of DNA.

d) These methods are limited by the difficulties in acquiring high labelling efficiency and the recycling of the label in the community resulting in a breakdown in selective enrichment

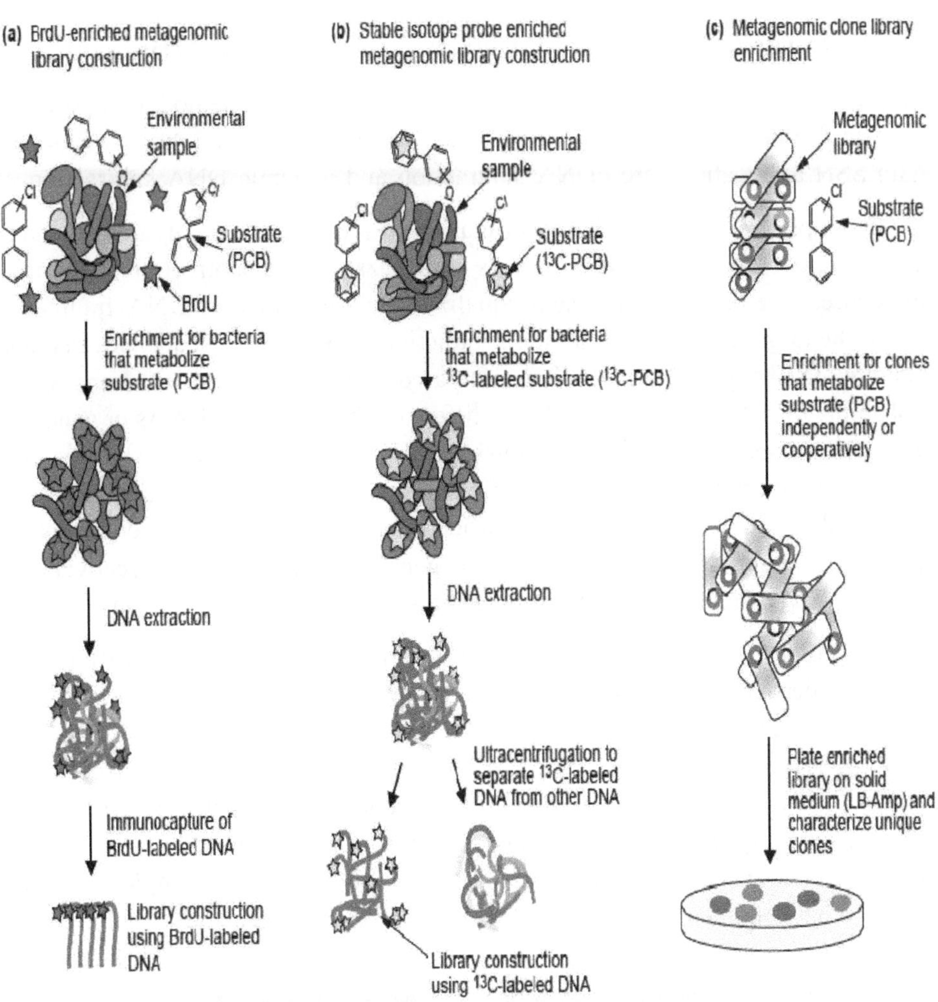

Fig. 6.1. Enrichment for specialized DNA from environmental samples using (a) BrdU-enrichment, (b) stable-isotope probing and (c) metagenomic library enrichment using PCBs as a model substrate.

6.3.1.3. *Mitigation of cross feeding by RNA-based SIP*

RNA-based SIP reduces the cross-feeding problems associated with DNA-based SIP. So far it is employed only to construct 16S rRNA gene libraries. If the isotope-labeled RNA presents a viable substrate for making cDNA, then libraries could be constructed from cDNA representing genes that are actively transcribed in the presence of the labeled compound. Metagenomic clone libraries could be screened in hybridization arrays using the cDNA clones as probes. This approach has attractive feature that one library constructed with non selected DNA from the entire community could be screened repeatedly using different probes, each developed from a distinct subset of the community

6.3.2. Suppressive subtractive hybridization (SSH)

Suppression subtractive hybridization (SSH) is a widely used method for separating DNA molecules that distinguish two closely related DNA samples. Two of the main SSH applications are cDNA subtraction and genomic DNA subtraction.

Principle of SSH: It is based primarily on a suppression polymerase chain reaction (PCR) technique and combines normalization and subtraction in a single procedure. The normalization step equalizes the abundance of DNA fragments within the target population, and the subtraction step excludes sequences that are common to the populations being compared. This dramatically increases the probability of obtaining low-abundance differentially expressed cDNAs or genomic DNA fragments and simplifies analysis of the subtracted library. The detailed principle is given in Fig 6.2. SSH technique is applicable to many comparative and functional genetic studies for the identification of disease, developmental, tissue specific, or other differentially expressed genes, as well as for the recovery of genomic DNA fragments distinguishing the samples under comparison.

Suppressive subtraction hybridization (SSH) identifies genetic differences between microorganisms and is therefore a powerful technique for specific gene enrichment. But the complexity of metagenomes makes this detection difficult. SSH has successfully been used on complex metagenomes and the sensitivity of the process can be increased by using multiple rounds of subtractive hybridization. Adaptors are ligated to the DNA populations and subtractive hybridization is carried out to select for DNA fragments unique to each DNA sample. This has typically been applied to analyse genetic differences between two closely related bacteria (e.g. in the identification of genetic elements contributing to pathogenesis).

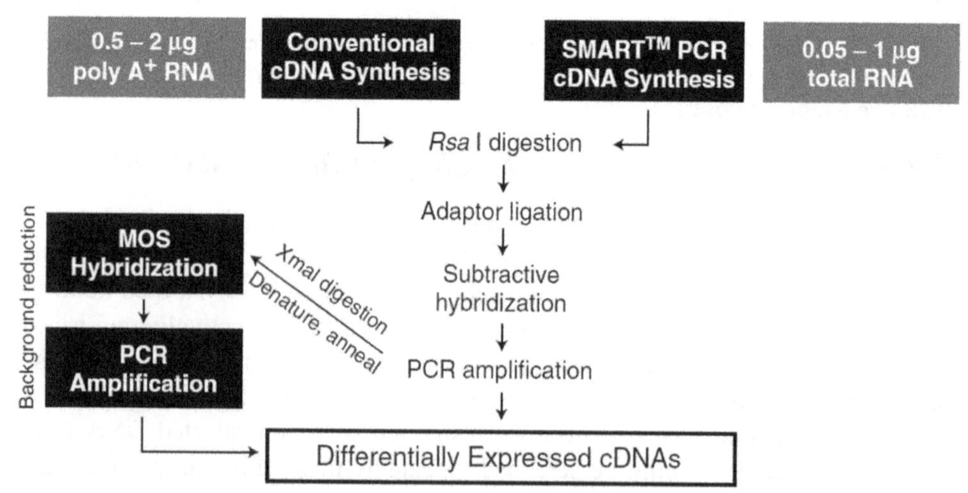

Fig 6.2. Brief overview of the suppression subtractive hybridization (SSH)

SSH technique is adapted to target specific genes in related metagenomes. For example, the identification of genes involved in the bioremediation of an environmental pollutant could be identified by comparison of a reference metagenome with a 'perturbed' metagenome (i.e. impacted by a specific pollutant). The relatively crude nature of this approach would only result in the identification of the total genetic difference between the two bacterial populations and would not be specific to genes of interest or genes whose expression was upregulated on addition of the environmental pollutant.

An especially challenging problem in SSH is the inclusion of "false positive" clones that generate a differential signal in a primary screening procedure, but are not confirmed by subsequent detailed analysis. To overcome this problem, a simple procedure called mirror orientation selection (MOS) can be used to substantially decrease the number of background clones. The MOS technique is based on the rationale that, after PCR amplification during SSH, each kind of background molecule has only one orientation relative to the adapter sequences.

6.3.3. Differential expression analysis (DEA)

Gene expression is a well coordinated system, and hence measurements on different genes are in general not independent. Fundamental to the task of analyzing gene expression data is the need to identify genes whose patterns of expression differ according to phenotype or experimental condition. Given more complete knowledge of the specific interactions and transcriptional controls, it is conceivable that meaningful comparisons between samples can be made by considering the joint distribution of specific sets of genes.

Principle: In current practice, differential expression analysis will therefore at least start with a gene-by-gene approach, ignoring the dependencies between genes. A simple approach is to select genes using a fold-change criterion. This may be the only possibility in cases where no, or very few replicates, are available. An analysis solely based on fold change however does not allow the assessment of significance of expression differences in the presence of biological and experimental variation, which may differ from gene to gene. This is the main reason for using statistical tests to assess differential expression. Generally, one might look at various properties of the distributions of a gene's expression levels under different conditions, though most often location parameters of these distributions, such as the mean or the median, are considered. One may distinguish between parametric tests, such as the *t*-test, and non-parametric tests, such as the Mann-Whitney test or permutation tests. Parametric tests usually have a higher power if the underlying model assumptions, such as normality in the case of the *t* test, are at least approximately fulfilled. Non-parametric tests do have the advantage of making less stringent assumptions on the data-generating distribution. In many microarray studies however, a small sample size leads to insufficient power for non-parametric tests. A pragmatic approach in these situations is to employ parametric tests, but to use the resulting *p*-values cautiously to rank genes by their evidence for differential expression.

To selectively enrich for a specific target gene within a metagenome a more practical approach would be to use one of several differential expression technologies that rely on the isolation of mRNA to target transcriptional differences in gene expression. DEA targets transcriptional differences in gene expression. Several innovative methods have been developed. These techniques have so far almost exclusively been used to study patterns in eukaryotic gene expression. Differential expression analysis (DEA) is a particularly effective enrichment tool. The expression profile of a culture grown from a metagenomic sample can be compared pre and post- exposure to a specific substrate or xenobiotic. In this way the expression of genes up-regulated for the specific activity can be identified. This type of approach was successfully applied to identify bacterial genes upregulated in the absence of iron.

Several variations in the basic concept exist. These include selective amplification *via* biotin and restriction-mediated enrichment (SABRE), integrated procedure for gene identification (IPGI), serial analysis of gene expression (SAGE), tandem arrayed ligation of expressed sequence tags (TALEST) and total gene expression analysis (TOGA). These techniques have been effectively applied for eukaryotic gene discovery. Their high sensitivity and selectivity should enable small differences in expression of single copy genes to be detected.

6.3.4. Affinity capture

Oligonucleotides covalently immobilised to a solid support can be used to affinity purify target genes. The slow kinetics of hybridization limit this process, but might be improved by using metagenomic mRNA or single-stranded DNA. This approach is still in development.

6.3.5. Microarray

Microarrays allow high-throughput robotic screening for targeting multiple gene products. The cost and availability of microarray technology is rapidly decreasing, making this an increasingly attractive option. In array-based differential expression analysis the problem is to generate a list of genes that are differentially expressed, being as complete as possible. The data is interpreted based on the differential gene expression related with statistical approaches (eg. T test).

6.3.6. Phage display

Phage-display expression libraries provide a means of isolating a given DNA sequence by affinity selection of the surface-displayed protein to an immobilised ligand. Biopanning involves repeated cycles of binding that will successively enrich the pool. After several rounds of enrichment, individual clones are characterised by DNA sequencing.This method is efficient and amenable to high-throughput screening, offering the potential to enrich even rare DNA sequences in the metagenome, but current phage technology limits expression of proteins >50kDa.

6.4. ENRICHMENTS FOR METAGENOMIC CLONES FORMING METABOLIC CONSORTIA (EXPRESSION SCREENING)

Most of the research on metagenomic libraries focuses on activities of individual clones. A key direction for development of this technology is to enrich for consortia of clones that together accomplish a desired function (Figure 6c). The approach is analogous to the isolation of bacteria that cooperate to degrade complex pools of polychlorinated biphenyl (PCB) compounds in soil, a process mediated by two or more species of bacteria in which each species contributes part of a 'mosaic pathway'. Such mosaic pathways indicate that it will not be possible to capture the genes for certain pathways on a contiguous piece of DNA; thus, approaches to study multiple clones simultaneously are essential to the long-term utility of metagenomics.

a) **Substrate-induced gene expression screening (SIGEX):** In 2005, Uchiyama and his co-workers introduced substrate-induced gene expression screening (SIGEX). This high-throughput screening approach employs an operon trap *gfp* expression vector in combination with fluorescence-activated cell sorting.The screen is based on the fact that catabolic-gene expression is induced mainly by specific substrates and is often controlled by regulatory elements located close to catabolic genes.

   To perform SIGEX, metagenomic DNA is cloned upstream of the *gfp* gene, thereby placing the expression of *gfp* under the control of promoters present in the metagenomic DNA. Clones influencing *gfp* expression upon addition of the substrate of interest are isolated by fluorescence-activated cell sorting. In this way, an aromatic- hydrocarbon-induced gene from a metagenomic library derived from groundwater was isolated. One drawback of this approach is the possible activation of transcriptional regulators by effectors other than the specific substrates.

b) **Metabolite-regulated expression (METREX):** Metabolite-regulated expression (METREX), is also a screening strategy designed by Williamson *et al.* 2005. To identify metagenomic clones producing small molecules, a biosensor that detects small diffusible signal molecules that induce quorum sensing is inside the same cell as the vector harboring a metagenomic DNA fragment. The main component of the biosensor is a quorum-sensing promoter which controls the reporter *gfp* gene. When a threshold concentration of the signal molecule encoded by the metagenomic DNA fragments is exceeded, green fluorescent protein (GFP) is produced. Subsequently, positive clones are identified by fluorescence microscopy.

   Recently, a new structural class of quorum-sensing inducers from the midgut microbiota of gypsy moth larvae by employing METREX was identified. A monooxygenase homolog which produced small molecules that induced the activities of *LuxR* from *Vibrio fischeri* and *Cvi*R from *Chromobacterium violaceum* was detected.

c) **Product induced gene expression (PIGEX):** In 2010, Uchiyama and Miyazaki introduced another screen based on induced gene expression, termed product induced gene expression (PIGEX). In this reporter assay system, enzymatic activities are also detected by the expression of *gfp*, which is triggered by product formation. In order to screen for amidases, the benzoate-responsive transcriptional regulator BenR is used as a sensor. Recombinant *E. coli* strains harboring the sensor and 96,000 metagenomic clones derived from activated sludge were co-cultured in microtiter plates in the presence of the substrate benzamide. In response to benzoate production by the metagenomic clones, the sensor cells fluoresce. In this way, three novel genes encoding amidases were identified.

Mini Quiz

1. What is the necessity for sample enrichment?
2. Explain stable isotope probing and its significance?
3. What are the limitations of BrdU enrichment?
4. Differentiate between DEA and SSH techniques?
5. Brief the enrichment techniques involved in expression screening?

CHAPTER - 7

METAGENOMIC PYROSEQUENCING AND MICROBIAL IDENTIFICATION

Metagenomics refers to culture-independent studies of the collective set of genomes of mixed microbial communities and applies to explorations of all microbial genomes in consortia that reside in environmental niches, in plants, or in animal hosts. Metagenomics and associated metastrategies have arrived at the forefront of biology primarily because of two major developments. i)The deployment of next generation DNA-sequencing technologies has greatly enhanced capabilities for sequencing large meta–data sets that were difficult to imagine just several years ago. ii) The second key development is an emerging appreciation for the importance of complex microbial communities in soil and involved in plant- microbe interactions, soil geochemical cycles and other process involved in sustainable maintenance of the ecosysytem.

The soil microbiome is the entire population of microbes that colonize the soil, including the rhizosphere rhizoplane, soil subsurface, vegetation and contaminated sites. The different microorganisms constituting the microbiome include bacteria, fungi and viruses. Depending on the context, parasites may also be considered to compose part of the indigenous microbiota. The "metagenome" of microbial communities is estimated to be approximately 100-fold greater in terms of gene content. These diverse and complex collections of genes encode a wide array of biochemical and physiological functions that may benefit the host, as well as neighboring microbes. Because bacteria form a predominant group of the microbiome and have the most comprehensively documented phylogenetic data sets and classification systems, focus is given to the bacterial community. Most of the data gathered to date have been compiled with Sanger (dideoxy) sequencing platforms, but this chapter throws light on the emerging parallel DNA-sequencing technologies based on pyrosequencing. Such next-generation sequencing systems introduce possibilities for deeply sequenced data collections and strategies aimed at microbial identification via single genetic targets or whole-genome methodologies.

7.1. MICROBIAL IDENTIFICATION IN METAGENOMICS – THE CRITICAL ISSUES

Several important issues have recently emerged with respect to metagenomics and microbes. One issue is that the science of metagenomics, in contrast to individual microbial or animal genomes, is ultracomplex and challenged by the existence of vast unknown or knowledge "deserts." Of the immense microbial taxonomic "space" in nature, only a restricted set of bacterial populations have been identified in the soil. The soil microbiota is a vast ecosystem but these estimates are in flux because the science of metagenomics and microbial pan-arrays is so new. A key remaining question is whether a core soil microbiome is definable Different soil types even in related and contiguous sites,may differ with respect to bacterial quantities and species composition. The depth of the soil, climate and nutrient availability also plays a major role in the microbial composition and dynamics. Bacterial species may not be randomly distributed in space or time. Healthy soil contains highly complex microbial populations, and the relative proportions of different bacteria may vary in different regions.

To address the central challenge of microbial identification in the context of mixed species communities requires refining the primary strategies for DNA sequencing– based bacterial identification. Prior studies of bacterial evolution and phylogenetics provided the foundation for subsequent applications of sequencing based on 16S rRNA genes (or 16S rDNA) for microbial identification. The initial studies were based on Sanger-sequencing strategies that included targeted sequencing of 16S rRNA genes (approximately 1.5 kb of target sequence). Such "long read" approaches enabled to identify many individual genera and species that could not be identified with biochemical methods. Sequence-based identification could be established with a reasonable amount of confidence from relatively long reads and with the aid of sequence classifier algorithms that included most of the 16S rDNA coding sequence; however, less than half of the coding sequence (approximately 500 bp), including several hypervariable regions, may be sufficient for genus- and species-level identification *via* Sanger sequencing.

7.2. PYROSEQUENCING – DIFFERENT APPROACH OF SEQUENCING

As sequence targets for microbial identification have become more precisely defined, the introduction of pyrosequencing has provided a user-friendly approach. Pyrosequencing has enabled more extensive sampling of microbial diversity with improved labor efficiencies. Specific genetic targets, such as hypervariable regions within bacterial 16S rRNA genes,may be amplified by the PCR and subjected to DNA pyrosequencing. DNA pyrosequencing, or sequencing by synthesis, was developed in the mid 1990s as a fundamentally different approach to DNA sequencing.

Principle: Sequencing by synthesis occurs by a DNA polymerase– driven generation of inorganic pyrophosphate, with the formation of ATP and ATP-dependent conversion of luciferin to oxyluciferin (Fig. 7.1). The generation of oxyluciferin causes the emission of light pulses, and the amplitude of each signal is directly related to the presence of one or more nucleosides. One important limitation of pyrosequencing is its relative inability to sequence longer stretches of DNA (sequences rarely exceed 100–200 bases with first- and second-generation pyrosequencing chemistries).

Based on pyrosequencing chemistry there are two sequencing platforms.

i) low-throughput sequencing platforms (e.g., the Biotage PSQ 96 System)
ii) high-throughput sequencing platforms (e.g., Roche/454 Life Sciences)

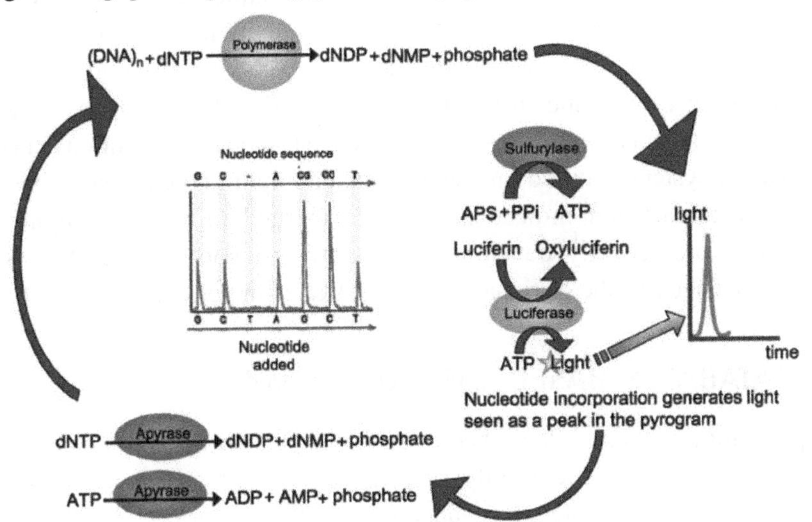

Fig 7.1. Pyrosequencing chemistry: biochemical reactions and chemicals involved in the generation of light signals by DNA pyrosequencing

Each peak in the pyrograms represents a pulse of light detected in the instrument. dNTP, deoxynucleoside triphosphate; dNDP, deoxynucleoside diphosphate; dNMP, deoxynucleoside monophosphate; PPi, pyrophosphate; APS, adenosine 5 -phosphosulfate. (Courtesy to: http://www.pyrosequencing.com)

7.3. APPLICATIONS OF DNA PYROSEQUENCING

i) DNA pyrosequencing has been successfully applied in a variety of applications, including genotyping, single-nucleotide polymorphism detection, and microorganism identification.
ii) Pyrosequencing has been used to detect point mutations in antimicrobial or antiviral resistance genes to explore the presence of drug-resistant microbes.
iii) The relatively short read lengths of DNA pyrosequencing have placed a premium on careful target selection and oligonucleotide primer placement.
iv) Pyrosequencing has been successfully applied to microbial identification by combining informative target selection (e.g., hypervariable regions within the 16S rRNA gene) and signature sequence matching.

Despite the fact that DNA pyrosequencing yields relatively short read lengths and limited amounts of sequence data per microbe, this strategy has been useful for microbial identification in different settings. Because of the relatively short read lengths, DNA pyrosequencing applications for microbial identification have focused attention on hypervariable regions within small ribosomal-subunit RNA genes, especially 16S rRNA genes. Once DNA sequence data are generated, sequences must be analyzed with special considerations in mind to facilitate accurate bacterial identification. First, different taxonomic classifications can be used for identification, and different species identifications may be generated, depending on the taxonomic scheme. The oldest and most traditional bacterial classification system is based on Bergey's taxonomy. In recent years phenotypic (e.g., biochemical) and molecular data are merged to create a higher-order taxonomy. More recently developed taxonomic schemes include the systems proposed by Pace and the National Center for Biotechnology Information (NCBI).

7.4. DATABASES BASED ON TAXONOMY

Multiple online databases have been developed on the basis of these different taxonomic schemes and currently provide convenient access to large rRNA sequence databases. Online rRNA databases include a variety of software tools for sequence classification and multiple sequence alignments for facilitating microbial identification. The ARB software package is a widely used program suite that includes open-source, directly interacting software tools that are linked to an integrated microbial sequence database. The most prominent databases include the

a) Ribosomal Database Project II (RDP II) (http://rdp.cme.msu.edu/):- RDP II is based on Bergey's taxonomy,which contains a relatively small number of phyla (divisions).

b) Greengenes (http://greengenes.lbl.gov):- Greengenes includes multiple taxonomic schemes, allowing the results of queries made with different classification schemes to be compared. It provides a 16S rRNA workbench for sequence-based microbial identification with different query and sequence-alignment tools. Green genes uses the NAST aligner tool and generates output that is compatible with ARB software tools so that different open-source environments may be linked via the Internet for comprehensive studies of microbial populations.

c) The ARB-SILVA:-It also offers a choice of microbial taxonomies, although it is more limited in its flexibility than Greengenes.

These software environments (Greengenes, RDP, ARB-SILVA) contain sequence-query tools, sequence-alignment programs, and sequence editors. Microbial identification depends on the taxonomic curation. The taxonomic schemes varies with respect to the number of phyla. Therefore, in addition to the routine issues of "splitting" and "lumping" taxa in the different schemes, one is confronted with different phyla (divisions) and different corresponding sub groupings (e.g., class, order, family).

7.4.1. Sequence classifier tools

Different supervised sequence classifier tools are available for matching test sequences with queried sequences. Compared with the Basic Local Alignment Search Tool (BLAST), supervised classifiers such as RDP Seq Match demonstrated greater accuracy in finding the most similar rDNA sequences.

i) Seq Match in RDP II : The RDP-based Seq Match *k*-nearest-neighbor (k-NN) classifier is effective at determining probable sequence identities on the basis of pairwise aligned distances. Alternatively, the RDP II group has developed its own naive Bayesian classifier that can be easily retrained as new sequences are incorporated into the rapidly expanding microbial-sequence databases. Seq Match in RDP II enable relatively short query sequences ≥ 50 bases in length to yield accurate microbial identifications.

ii) Bayesian classifier: This tool was also developed by RDP group. It uses information averaged within the entire genus and is less influenced by individual misplaced sequences.

Despite the use of 2 supervised classifiers with the same database, different results can be generated for particular sequences, particularly with phylogenetically broad genera such as *Clostridium*.

7.5. NEXT GENERATION DNA SEQUENCING TECHNOLOGIES—METAGENOMIC PYROSEQUENCING

Until recently, Sanger-sequencing methods were primarily used to generate data in most microbial genome and metagenomics sequencing projects. The rapid development of parallel, high-throughput sequencing technologies during the current decade has led to commercialization and widespread adoption of next-generation sequencing technologies. In contrast to a relatively homogeneous DNA-sequencing enterprise in the 1990s, current large-scale genome and metagenome sequencing projects are deploying multiple platforms and different sequencing chemistries in parallel. As of June 2008, 3 leading vendors of nextgeneration platforms commercially distribute machines for high-throughput sequencing: Roche/454 Life Sciences, Illumina/Solexa, and Applied Biosystems (SOLiD). Different generations of the machines have been created, with different levels of performance.

454 Life Sciences (now a subsidiary of RocheDiagnostics) was the one company that commercially developed pyrosequencing for metagenomics. With respect to 454 sequencing, third-generation platforms that provide longer read lengths are now emerging. The first-generation instrument GS 20, yielded 100-bp reads and 30–60 Mb per run. The second- and third-generation instruments include the FLX (now GS FLX Standard) and XLR (now GS FLX Titanium) platforms, respectively. The FLX was released in 2006 and yielded 250-bp reads and approximately 150 Mb/run. The XLR, released in 2008, yields demonstrably higher read lengths, exceeding 350 bp and approximately 400 Mb/run. The 454 instruments are widely deployed next-generation sequencing systems currently in the scientific community, and these pyrosequencing-based platforms preceded other high throughput platforms, such as the Illumina/Solexa and SOLi Dtechnologies mentioned above.

7.5.1. The 454 technology

Each 454 platform uses a modern adaptation of DNA-pyrosequencing chemistry. The 454 technology is highly advantageous because of the technical robustness of the chemistry. The relatively long reads generated by 454 sequencing allow more frequent unambiguous mapping to complex targets than the products of the other next-generation technologies, which feature shorter reads. During the past decade, sequencing read lengths have improved because of refinements in pyrosequencing biochemistry, such as the addition of recombinant enzymes including single-stranded binding protein. Advances in microfluidics technologies within instruments have increased the speed of sequencing reaction cycles so that more cycles can be performed per unit time in second- and third generation sequencers. Additionally, the large numbers of reads per run that are possible with 454 technology deliver much greater depth of coverage for metagenomic sequencing than Sanger sequencing.

Accurate, proofreading, and thermostable DNA polymerases and the application of temperature gradients during PCR amplification represent key considerations for maximizing the specificity of DNA amplification prior to 454 sequencing. Improving the accuracy of target amplification can minimize subsequent errors produced in high-throughput pyrosequencing.

7.6. METAGENOMICS: SEQUENCING OF 16S RDNA AMPLICONS

Metagenomics strategies may be directed at examining microbial composition or the broader issue of tackling phylogenomic diversity of highly complex microbial populations. One basic approach is to identify microbes in a complex community by exploiting universal and conserved targets, such as rRNA genes. By amplifying selected target regions within 16S rRNA genes (Fig. 7.2), microbes (specifically bacteria and archaea) can be identified by the effective combination of conserved primer-binding sites and intervening variable sequences that facilitate genus and species identification (Fig. 7.3). The 16S rRNA gene in bacteria consists of conserved sequences interspersed with variable sequences that include 9 hypervariable regions (V1–V9, Fig.7. 2). The lengths of these hypervariable regions range from approximately 50 bases to 100 bases, and the sequences differ with respect to variation and in their corresponding utility for universal microbial identification. Reads obtained by 454 sequencing encompass multiple hypervariable regions with the second-generation platforms such as the FLX. Third generation 454 sequencing platforms such as the LXR will generate reads exceeding 350 bp and further facilitate the sequencing of multiple hypervariable regions.

Enormous challenges are left with any metagenomics strategy for a highly conserved gene. A parallel analysis of 3 different hypervariable regions of 16S rDNA sequence (V2–V3, V4–V5, and V6–V8 regions) was effective in determining the composition of bacterial consortia in maize rhizospheres. So there exists variability in the representation of operational taxonomic units (OTUs), which depends on the hyper variable region.

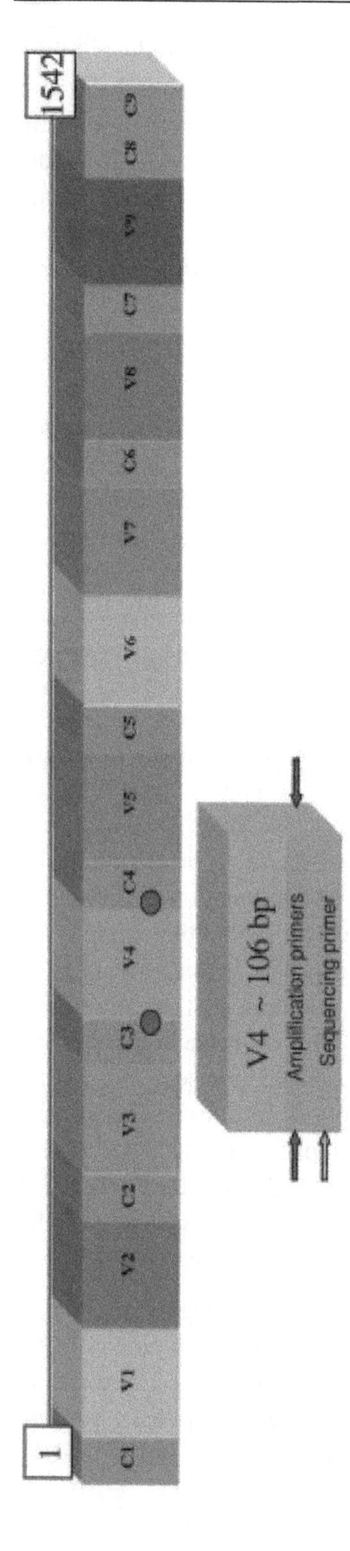

Fig. 7.2. Conserved and hypervariable regions in the 16S rRNA gene. The interspersed conserved regions (C1–C9) are shown in gray, and the hypervariable regions (V1–V9) are depicted in different colors. Also illustrated is an example of primer selection for DNA amplification and sequencing-based microbial identification (V4 subregion with pink circles and arrows representing primer-binding sites)

7.6.1. Whole-genome shortgun sequencing strategies

Microbial 16S rDNA sequencing is considered the gold standard for characterization of microbial communities, but 16S rDNA sequencing may not be sufficiently sensitive for comprehensive microbiome studies. rRNA gene– based sequencing can detect the predominant members of the community, but these approaches may not detect the rare members of a community with divergent target sequences. Primer bias and the low depth of sampling account for some of these limitations, which could be improved with 454 sequencing of entire microbial genomes. To overcome the limitations of single gene– based amplicon sequencing by pyrosequencing, whole-genome shotgun sequencing has emerged as an attractive strategy for assessing complex microbial diversity in mixed populations.

Whole genome–based approaches offer the promise of more comprehensive coverage by high throughput, parallel DNA-sequencing platforms, because they are not limited by sequence conservation or primer-binding site variation within a specific target (Fig.7.3). Whole-genome approaches enable to identify and annotate

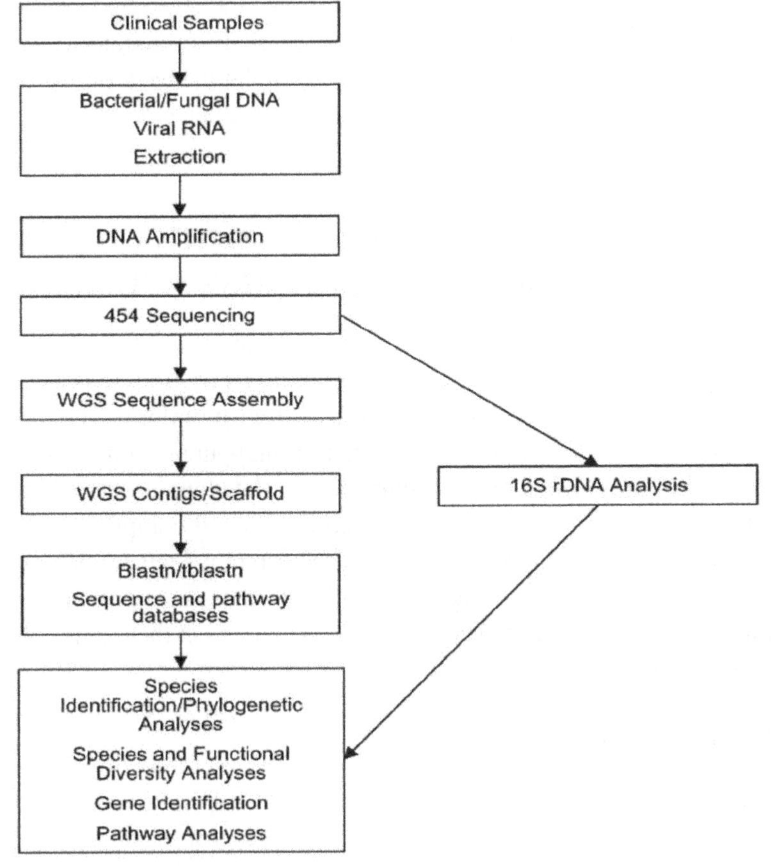

Fig. 7.3. 454 sequencing technology for metagenomics

diverse arrays of microbial genes that encode many different biochemical or metabolic functions. Novel genes and functions are being discovered because of the massive data sets obtained in whole-genome shotgun sequencing of marine samples. The assessment of aggregate biological functions or community phenotypes based on functional metagenomics may depend on whole-genome metagenomic sequencing strategies. Arguably, whole genome approaches provide the only bona fide strategies for true metagenomics studies.

7.6.2. Limitations of whole-genome strategies

The challenges and limitations of whole-genome strategies are:

i) Requires relatively large amount of starting material
ii) Potential contamination of metagenomic samples with host genetic material
iii) High numbers of genes of unknown function or lacking quality annotation

The sequence bias can be minimized with polymerases, such as the Phi 29 DNA polymerase. In addition to the possibility of amplification bias, the potential of WGA to co-amplify contaminating (host) DNA poses a significant challenge, and such host DNA co-amplification may overwhelm the bacterial DNA sequence data in the sample. Different subtraction strategies are being developed to minimize this possible barrier.

7.7. NEXT-GENERATION MICROBIAL-IDENTIFICATION STRATEGIES: METAGENOMICS AND INFORMATICS

The primary challenge for metagenomics studies at the analytical end is how to obtain accurate microbial identification for hundreds or thousands of species in a reasonable time and for a reasonable cost. Current bioinformatics throughput is too slow and not sufficiently automated. High-throughput methods of metagenomics rDNA analyses are needed and are currently in development. Clearly, sufficient computational power is necessary, although distributed computing networks and robust server technology may eventually meet current metagenomics data-analysis demands in research settings. Beginning with sequence collection and verification, algorithms must be in place to trim sequences and to meet the quality of individual reads *via* various strategies.

7.7.1.Sequence trimming

Sequence trimming is the removal of primer and low-quality sequence data before sequence assembly. Once the sequence reads have been trimmed sequences can be aligned with multiple sequence alignment programs such as NAST or MUSCLE.

Another problem is that the PCR may generate sequence chimeras because of errors that couple disparate DNA sequences during the amplification process. Chimera-checking software has been developed so that amplicons can be vetted for the presence of "sequence hybrids" in software environments such as Green genes and RDP and with tools such as Bellerophon or Pintail.

Once high-quality sequences have been obtained from mixed species communities, the next challenge is to accurately identify many microbes in parallel. Sequences can be identified with facile classifiers such as the Bayesian Classifier in the RDP system and can be compared with robust multisequence alignment programs such as NAST or MUSCLE. Existing software environments such as RDP, Greengenes, or ARB-SILVA include multisequence alignment programs that can be effectively coupled with sequence editors in an integrated fashion.

7.7.2. Sequence binning

For large data sets, 16S rDNA sequences may be binned with programs such as Fast GroupII, and tallies of OTUs may be generated from these bins. Aligned sequences may also be classified against databases such as prokMSA (prokaryotic multiple sequence alignment) in Greengenes, and tallies of phyla may be examined and ultimately displayed as relative abundance histograms so that differences in proportions of different bacterial groups can be compared.

7.7.3. Phylogenomics

Novel informatics approaches such as CARMA enable sequences encoding protein segments as short as 27 amino acid residues from whole-genome sequencing projects to be applied in microbial identification strategies for comparative metagenomics. High-throughput informatics approaches must be developed to cope with the demands of next generation DNA sequencing. One new strategy, automated simultaneous analysis phylogenetics (ASAP), offers an automated strategy for phylogenomics that may facilitate analyses of high volumes of sequence data, especially in whole genome–based microbiome explorations. In addition to accurate microbial identification, indices and algorithms have been developed to assess microbial diversity in the context of the microbiome.

Phylogenetic distance matrices are constructed in programs such as DNAML. Distance matrices can be transferred to DOTUR for construction of collector's curves, rarefaction curves, calculations of Chao and ACE richness estimates, and computations of Simpson and Shannon indices of diversity.

7.8. SPECIAL CHALLENGES OF WHOLE GENOME SEQUENCING IN FUNGAL AND VIRAL METAGENOMICS

Whole-genome sequencing of metagenomic samples is likely to reveal many bacteriophage, prophage, and eukaryotic viral sequences, but viral metagenomics analyses show only 60% unique sequences thus representing unknown viral species. As such, viral sequences may be missed by whole-genome sequencing. Preparation of viral nucleic acids may include the filtration of samples to remove host and bacterial cells, followed by treatment of the filtrate with nucleases to remove host nucleic acids. Such virome sequencing strategies could easily be adapted to high-throughput 454 sequencing platforms. In the area of eukaryotic metagenomics, limited studies have been performed on fungal diversity in soil and fungi associated with plants. The internal transcribed spacer regions downstream of 18S rRNA genes may be useful for fungal identification.

So from this chapter it is very well understood that the science of metagenomics is currently in its pioneering stages of development as a field, and many tools and technologies are undergoing rapid evolution. In addition to paradigmatic shifts toward next generation DNA-sequencing technology based on novel chemistries, bioinformatics tools are also being redefined in fundamental ways to accommodate the large volumes of data. In addition to massive data sets, new questions are being posited that challenge the abilities of current algorithms to deliver meaningful answers in the context of biology. The open-source software movement and "wikinomics," or mass collaboration approaches in biology, have already established a foundation for the metagenomics arena with software environments such as ARB. Complementary strategies for microbial identification that depend on pan-microbial microarrays with known sequences, such as the Phylochip or the Virochip are available. In future sequence polymorphisms and implied biological functions will be considered for exploring the microbial communities.

Mini Quiz

1. Discuss the critical issues in microbial identification by metagenomics?
2. Explain the principle of pyrosequencing or sequencing by synthesis?
3. What is Bayesian classifier?
4. Write about the conserved and hypervariable regions in 16S rRNA?
5. Differentiate between sequence trimming and sequence binning?

CHAPTER - 8

METAGENOMIC GENE DISCOVERY

The recent development of technologies designed to access the wealth of genetic information through environmental nucleic acid extraction has provided a means of avoiding the limitations of culture-dependent genetic exploitation. After biotope selection and sample or culture enrichment (if desired), nucleic acid is extracted from the environmental sample. The approach might involve metagenomics (environmental genomic DNA) or metatranscriptomics (environmental mRNA reversed transcribed to complementary DNA, cDNA) and an enrichment or selection can be applied.

Gene enrichment selects for differentially expressed genes using techniques such as differential expression analysis (DEA) and gene targeting. Genome enrichment uses techniques such as stable isotope probing (SIP), 5'Bromo-2-deoxyuridine (BrdU)-labelling and suppressive subtractive hybridization (SSH) to enrich or select for genomes of interest. Downstream screening approaches can be activity-based through the screening of expression libraries, sequence-dependent by using gene targeting or can be sequence-independent through the direct sequencing of the metagenome. The final expression requires a full-length open reading frame (ORF) expressed in a suitable host to generate a functional gene product (Fig 8.1).

The isolation of metagenomic DNA, library construction, gene and genome enrichment strategies were discussed in chapters 4, 5 and 6.

8.1. SEQUENCE BASED SCREENING

8.1.1. Gene targeting

Gene-specific PCR has been used extensively to probe communities for microorganisms with specific metabolic or biodegradative capabilities. For example, the targeting of genes such as methane monooxygenase, methanol dehydrogenase and ammonia monooxygenase was used to identify methanotrophic and chemolithotrophic ammonium-oxidizing bacteria. The biodegradative potential of indigenous microbial populations has been assessed by screening metagenomic extracts for the presence of catechol 2,3-dioxygenase, chlorocatechol dioxygenase and phenol hydroxylase genes. Other reported examples include the identification of denitrifying bacteria and polyhydroxyalkanoate producing bacteria.

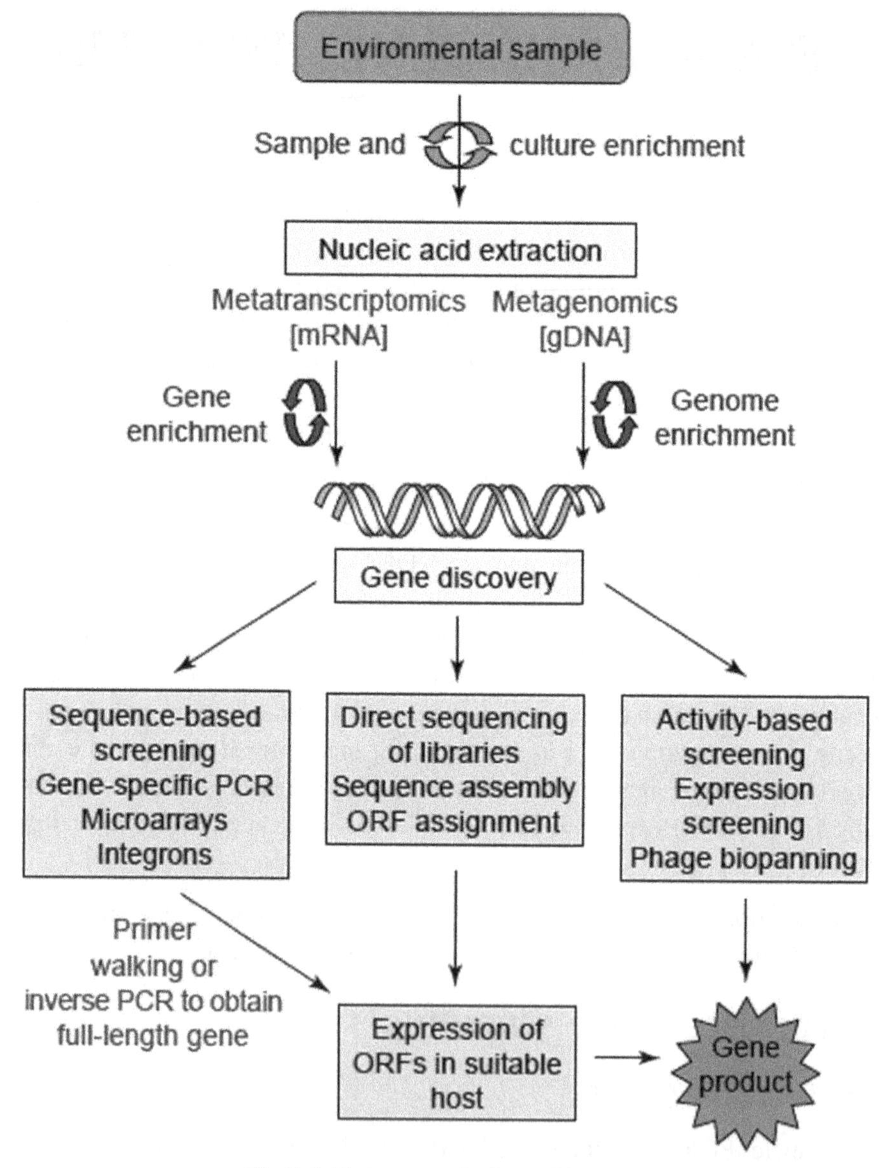

Fig 8.1. Metagenomic Gene Discovery

However, as a tool for biocatalyst discovery, gene-specific PCR has two major drawbacks.

i) the design of primers is dependent on existing sequence information and skews the search in favour of known sequence types. Functionally similar genes resulting from convergent evolution are not likely to be detected by a single gene-family-specific set of PCR primers.

ii) only a fragment of a structural gene will typically be amplified by gene-specific PCR, requiring additional steps to access the full-length genes.

Amplicons can be labelled as probes to identify the putative full-length gene(s) in conventional metagenomic libraries. Alternatively, PCR-based strategies for the recovery of either the up- or down-stream flanking regions including universal fast walking, cassette PCR panhandle PCR, random primed PCR, inverse PCR and adaptor ligation PCR can be used to access the full-length gene. These methods are technically more difficult to apply at a metagenomic level owing to the increased complexity of a metagenomic DNA sample.

8.1.1.1. *Novel genes recovered by gene specific PCR based strategies*

PCR based strategies have been used successfully for the recovery of novel gene variants of 2,5 diketo-D-gluconic acid reductase from environmental DNA. Because these approaches can be laborious and time-consuming, innovative alternatives have been developed. Cassette PCR has been used to isolate the central fragment of catechol 2,3-dioxygenase genes from genomic DNA obtained from a phenol and crude oil-degrading bacterial consortium. The internal fragment of a previously cloned full-length copy of a catechol 2,3-dioxygenase gene was then replaced with the PCR-derived internal fragment, thus constructing a novel hybrid catechol 2,3- dioxygenase gene. This approach can be combined with PCR mutagenesis and/or chimeragenesis to generate highly diverse protein variants incorporating random and directed molecular evolution.

8.1.2. RNA as an effective biomarker for gene discovery – the RT-PCR

The use of RNA might be more effective than DNA for profiling functional microbial communities because RNA is a more sensitive biomarker owing to its high turnover. Reverse transcriptase PCR (RT–PCR) has been used to recover genes from environmental samples, for example in the isolation of naphthalene-degrading enzymes from microorganisms present in a coal tar waste.

8.1.2.1. *Merits and Demerits*

This approach suffers technical difficulties associated with mRNA recovery from environmental samples. But it benefits from wider genomic access (includes structural genes from lower eukaryotes as well as from prokaryotes) and the facility to select for functional genes in response to alterations in environmental conditions.

8.1.3. Integrons

Integrons are naturally occurring gene capture, dissemination and expression systems that have until recently primarily been associated with antibiotic resistant and pathogenic bacteria. They are widely dispersed in nature and could play a significant role in bacterial genome evolution.

8.1.3.1. *Features of an integron*

The key structural features of an integron include a gene cassette integration site (*att1*), an i*ntI* gene that encodes an integrase and two promoters that drive the expression of the integrase gene and the incorporated gene cassettes. The mobile element in the system is the gene cassette, which consists of one or more open reading frame(s) (ORFs) and associated chromosomal attachment sites *(attC*, also referred to as the 59 base elements). The integrase catalyses the insertion of the gene cassette into the integration site controlled by the strong promoter via site-specific recombination using *att1* and *attC* as its substrates. Integrons therefore act as a repository of ORFs coding for many gene products and potentially provide a source of novel genes.

Primers designed to target the conserved regions within the 59 base elements have successfully been used to recover novel genes homologous to DNA glycosylase, phosphotransferase, methyl transferase and thiotransferase. The specificity of this system for gene targets could be improved by using a primer specific for the gene of interest and one targeting a flanking 59 base element.

8.1.4. Homologous recombination cloning

Homologous recombination cloning can be used for single-step gene targeting and screening with only those recombinants containing the gene of interest viable after transformation. This method requires the design of an *E. coli* host containing a vector DNA sequence homologous to the 50- and 30- sequences flanking the gene of interest. The efficiency of bacterial homologous recombination has been improved and commercial systems are now available (eg. www.genebridges.com). So far, homologous recombination cloning has not yet been applied to metagenomic gene discovery.

8.1.5. Affinity binding

Methods requiring only one gene-specific primer impose less sequence-dependent bias compared with standard twin-primer PCR amplification procedures. An elegant application of this method would be the use of immobilised oligonucleotides designed to target a specific gene fragment or consensus sequences by affinitybinding. This approach is, of course, used routinely for recovery of polyA RNA cDNA library construction, but has not been applied to gene targeting.

Affinity capture should be equally applicable to either denatured cDNA or genomic DNA fragments and yields could be further enhanced with prior linking of adaptors so that affinity selected DNA fragments could be PCR amplified using linker-specific primers.

8.1.6. Microarrays

Microarrays represent a powerful high-throughput system for analysis of genes. They are typically used to monitor differential gene expression, to quantify the environmental bacterial diversity and catalogue genes involved in key processes. Microarrays of immobilized oligonucleotide gene targets have also been used to select appropriate biotope samples for metagenomic library screening. Such arrays could also be used for the affinity capture of targets as a means of enrichment before construction of metagenomic libraries. Microarray technology could also be used for the pre-selection of genes in metagenomic libraries before shotgun sequencing, thereby reducing the sequencing burden and reducing the proportion of sequences unassigned by database sequence similarity searches.

8.1.7. Primer walking or inverse PCR

Primer walking is a step-by-step approach to sequencing long DNA templates from end to end that overcomes the inability of the Sanger chain termination method to read more than a few hundred bases in a single reaction. After an initial round of sequencing from a known sequence at one end of the template, each subsequent round is initiated from a new primer, which is based on the end of the sequence obtained from the previous reaction.

8.2. METAGENOMIC DNA LIBRARIES

The basic steps of DNA library construction (generation of suitably sized DNA fragments, cloning of fragments into an appropriate vector and screening for the gene of interest) have been extensively discussed in previous chapters. As there are no obvious limitations in translating the technologies of genomic library construction and screening to metagenomic libraries, it is perhaps surprising that metagenomics only developed in the mid 1990s with the successful application of library construction to marine metagenomes.

Subsequent metagenomic gene mining work by Recombinant Biocatalysis Ltd (now Diversa Corporation) and several other laboratories demonstrated the successful recovery of novel genes from metagenomic gene expression libraries. The approach taken by each has been broadly similar,although a variety of vector and host systems have been used (Table 8. 1). Functional expression is commonly used as a method to screen for specific gene classes. However, such libraries are amenable to screening by virtually any method that can be adapted to deal with large clone populations.

DNA fragmentation is a significant problem when constructing metagenomic libraries. The vigorous extraction methods required for high yields of DNA from environmental samples often result in excessive DNA shearing. This precludes the construction of libraries using cohesive ends because highly sheared DNA (e.g. 0.5–5 Kbp fragments) cannot be restricted to generate ligatable sticky ends without significant loss of the total gene complement. An alternative approach uses blunt-end or T–A ligation to clone randomly sheared metagenomic fragments.

Cosmid and bacterial artificial chromosome (BAC) libraries have been widely used for the construction of metagenomic libraries. The ability to clone large fragments of metagenomic DNA allows entire functional operons to be targeted with the possibility of recovering entire metabolic pathways. This approach has successfully been applied to the isolation of several multigenic pathways such as that responsible for the synthesis of the antibiotic violacein. Fosmid vectors provide an improved method for cloning and stably maintaining cosmid-sized (35–45 Kbp) inserts in *E. coli*.

Phage-display expression libraries provide a means for isolating DNA sequences by affinity selection of the surface-displayed expression product. This method is efficient and amenable to high-throughput screening, offering the potential to enrich even rare DNA sequences in the metagenome. However, phage display is limited by the expression capacity of the bacteriophage, a protein size upper limit of around 50kDa.

The limitation of *E. coli* as a host for comprehensive mining of metagenomic samples is highlighted by the low number of positive clones obtained during a single round of screening (typically less than 0.01%). A recent *in silico* study indicates that it is virtually impossible to recover translational fusion products owing to the high number of clones (>10^7) that would need to be screened. Intuitively, expression from native promoters and read-through transcription from the vector-based promoter offer the best chance for recovery of heterologously expressing genes. Statistically, for a small insert (<10 Kbp) library, between 10^5 and 10^6 clones need to be screened for a single hit. This suggests that without sample enrichment the discovery of specific genes in a complex metagenome is technically challenging. The assumption that expression in an *E. coli* host will not impose a further bias is largely untested. Although the *E. coli* transcriptional machinery is known to be relatively promiscuous in recognizing foreign expression signals, a bias in favour of Firmicutes genes has been established. The further development of host screening systems is therefore a fruitful approach for the more effective future exploitation of metagenomes.

8.3. METAGENOMIC cDNA (TRANSCRIPTOMIC) LIBRARIES

Owing to the presence of intronic sequences, metagenomic expression libraries are generally not suitable for mining eukaryotic genes. The large-scale sequencing of clones from cDNA libraries has long been a rapid means of discovering novel eukaryotic genes. Acknowledging the technical difficulties of metagenomic mRNA isolation, there is no inherent reason why these technologies cannot be applied to exploit unculturable eukaryotic enzyme genomes via the construction of metagenomic cDNA libraries. Metagenomic cDNA libraries cannot be as comprehensive as genomic libraries because they can never represent non expressed genes. In addition, the process of RT–PCR amplification limits the size of inserts and could impose a large sequence-dependent bias on the library.

8.4. METAGENOME SEQUENCING

The sequencing and analysis of large fragments of genomic DNA from uncultured microorganisms are well established technologies. These studies have laid the groundwork for the ultimate in metagenomic gene discovery – the sequencing of complete metagenomes. With the relatively recent advent of automated, high throughput sequencing facilities and of powerful algorithms for sequence assembly, these projects are now technically feasible, albeit financially ambitious. The scale of the task is not trivial – a gram of soil or litre of seawater contains many

Table 8.1. Characteristics of metagenomic libraries. Examples of libraries constructed for gene targeting are shown[a]

Target gene	Host or vector systems used	Library size (number of independent clones)	Average insert size or range of size (Kbp)	% Prokaryote metagenome represented[e]	References
Chitinase	Lambda Zap II/ GigapackIII	750 000[b]	2–10	11[c,d]	Cotrell *et al.*, 1999
4-hydroxybutyrate dehydrogenase; lipase, esterase; cation/H^+ antiporters	E. coli DH5a/p Bluescript	930000	5–8	14[d]	Henne *et al.*, 1999, 2000. Majernik *et al.*, 2001
Lipase, amylase,	E. coli DH10B/	3648	27	0.2[d]	Rondon *et a l.*,2000
nuclease	pBeloBAC11	24576	44.5	3[d]	Rondon *et a l.*,2000
Heme biosynthesis; phosphodiesterase	E. coli TOP10/p CR-XL-TOPO	37000	1–10	0.5[d]	Wilkinson *et al.*, 2002
Polyketide biosynthesis	E. coli, S. lividans shuttle cosmid	5000	50	0.7	Courtois *et al.*, 2003
Alcohol oxidoreductase	E. coli/ pSKC	360 000 583 000 324 000	4.4 3.8 3.5	5 3 2	Knietsch *et al.*,2003

[a] Caution is advised in attempting to directly compare metagenomic libraries made in different laboratories using different systems.

[b] Number of clones screened.

[c] 1800 genomic species were estimated for an oligotrophic open ocean environment. Owing to the coastal location of the sample used in this study , we are assuming a 10-fold higher species diversity.

[d] In making these calculations, we have assumed an average of 104 prokaryotic species per environmental sample and an average prokaryotic genome size of 4Mbp.

[e] Chemical lysis methods of DNA extraction from soil samples are relatively non-aggressive and we assume that the contribution from eukaryotic (particularly fungal) genomes is minor. We acknowledge that this assumption might be invalid.

thousands of unique viral and prokaryotic genomes, hundreds of lower eukaryote species and DNA derived from higher eukaryotes. Using conservative estimates of genome sizes, soil metagenomes could constitute between 20 and 2000 Gbp of DNA sequences.

The sequencing of 76 Mbp of DNA from an acid mine drainage biofilm was the first reported study of this kind. The low biodiversity of the sample enabled the shotgun sequence assembly of two complete genomes. More than 4000 putative genes were identified, thereby providing insight to the metabolic pathways of the biofilm community. The sequencing of the Sargasso Sea metagenome was more challenging with the sequencing of >1 Gbp of DNA. Approximately 1.2 million putative genes were identified, clearly illustrating the enormous power of this approach for gene discovery. However, the functional assignment of novel genes (*i.e.* those with no database homologue) is in a state of infancy, with 'evidence-based' gene finder programs having limited success. The high biodiversity of the Sargasso Sea and poor sequencing coverage enabled the assembly of only two near-complete genomes.

Whole genome assembly could be improved by normalising abundant sequences using a combination of small, medium and large insert libraries and by increasing the coverage of sequencing (at a cost). Recently, differences in tetranucleotide repeat numbers between genomes have proven useful tools for discrimination, provided that the sample is low in complexity and the genomes are equally represented. However,nucleotide polymorphisms, gene rearrangements, gene duplications and horizontal gene transfer are all factors that will impact on reliable genome assembly. Eukaryotic metagenome sequencing poses even greater challenges owing to the presence of larger genome sizes,introns and 'junk' DNA. The use of metatranscriptomics and cDNA libraries might, to some extent, overcome these limitations.

8.5. COMMERCIAL SUCCESSES

Since the inception of two pioneering commercial metagenomics ventures in the late 1990s (Recombinant Biocatalysis Ltd of La Jolla and TerraGen Discovery Inc. of Vancouver; www.cubist.com) these technologies have been taken up by several of the biotechnology giants, and have been the focal area of several start-up companies (see Table 8.2). Recombinant Biocatalysis Ltd, now Diversa Corporation; www.diversa.com), is the acknowledged leader in the field with impressive lists of libraries derived from global biotopes and of cloned enzymes in a range of enzyme classes. Several other smaller biotechnology companies appear to be competing in the same market sector. The relatively small size of

Table 8.2. Commercialization of metagenomic technologies[a]

Company	Target products	Classes	Products and market	Commercial interest
BASF www.corporate. basf.com	Enzymes	Amylase Hydratase	Acidophilic glucoamylase	Food industry, aiding with the digestion of starch
Bioresearch Italia, SpA (Italy)	Anti-infectives	N.D.	Dalbavancin	Development of human gene targeted therapeutics and novel anti-infective
B.R.A.I. Nwww.brain-biotech.de	Bioactive peptides and enzymes for pharmaceuticals and agrochemicals	N.D.	Nitrile hydratases Cellulases	Degussa AG Partnership for the industrial processes
Cubist pharmaceutical shttp://www.cubist.com/	Anti-infectives	N.D	N.D.	Various commercial relationships. Variety of products in Stage I, II and III trials
Diversa www.diversa.com	Enzymes	Nitrilase Glycosidase Phytase	Discovery of> 100 novel nitrilases, Production of Lipitor Pyrolasee 160 and Pyrolasee 200; Phyzymee XP	Drug, lowering cholesterol levels Broad spectrum b-mannanase and b-glucanase added to animal feed to break down indigestible phytate in grains and oil seeds to release digestible

[Table Contd.

Contd. Table]

Company	Target products	Classes	Products and market	Commercial interest
	Biometabolites	Fluorescentprotein	DiscoveryPointe Green-FP* and Cyan-FP*	Novel green and cyan fluorescent proteins for potential use in drug discovery, commercial screening and academic research
Diversa and Invitrogen www.invitrogen.com	Enzymes	DNA polymerase	Thermal Acee and Replicasee DNA for research and diagnostics	Research and diagnostics
EMetagen www.emetagen.com	Enzymes; antibiotics; small active molecules	Polyketides	eMetagen Gene and Pathway, Banks e Large clone DNA libraries encoding biosynthetic pathways for5000 to 20 000 secondary metabolites	Food, agriculture, research and other commercial applications Pharmaceuticals: antimicrobial, anticancer and other bioactive properties
Kosan Technology www.kosan.com	Antibiotics	Polyketides	Adriamycin, Erythromycin, Mevacor,Rapamycin, Tacrolimus(FK506), Tetracycline, Rapamycin,	Therapeutic drugs
Genencor www.genencor.com	Enzymes	LipaseProtease	Washing powder and alkaline tolerant protease.	Cleaning industry

Table Contd.

Contd. Table]

Company	Target products	Classes	Products and market	Commercial interest
Libragen www.libragen.com	Antibiotics and biocatalysis for pharmaceuticals	N.D	Anti-infective and antibiotic discovery Biocatalysis discovery for pharmaceuticals (partnership with Synkem)	Medicine; synthesis of pharmaceuticals
Prokaria www.prokaria.is/	Enzymes	Rhamnosidase b-1,6 Gluconase Single stranded DNA ligase	Food and agricultural industry	Food industry
Proteus www.proteus.fr	Enzymes; antibiotics; antigens	Not specified	Research and diagnostics Anti- phytopathogenic fungal agent	Products for the agricultural, environmental, food, medical and chemical industries Development of novel biomolecules
Xanagen www.xanagen.com	Libraries	Gene products	Unspecified	Services in library construction, screening and annotation

[a]Note: Some of the products listed above may have been derived from metagenomic libraries with prior enrichments or from single genomes N.D. - no details available or products still under development.

the industrial enzyme market compared with the pharmaceuticals market suggests that a switch in product focus might not be unexpected. Although the authors are unaware of any successfully commercialised therapeutics derived from metagenomic screening programs, the normal timelines for the identification, development, evaluation and approval of products for the pharmaceutics market are longer than the existence of metagenomics as a research field.

Mini Quiz

1. Why RNA might be more effective than DNA for profiling microbial communities?
2. What are integrons?
3. Define inverse PCR?
4. What is the advantage of phage display expression libraries?
5. Mention few novel genes recovered by gene specific PCR?

CHAPTER - 9

TAXONOMIC AND FUNCTIONAL DIVERSITY OF MICROBIAL COMMUNITIES

Initially, metagenomics was used mainly to recover novel biomolecules from environmental microbial assemblages. The development of next-generation sequencing techniques and other affordable methods allowing large-scale analysis of microbial communities resulted in novel applications, such as comparative community metagenomics, metatranscriptomics, and metaproteomics. The combination of DNA-based, mRNA-based, and protein-based analyses of microbial communities present in different environments is a way to elucidate the compositions, functions, and interactions of microbial communities and to link these to environmental processes.

9.1. MICROBIAL DIVERSITY ANALYSIS

Microbial diversity in environments such as soil, sediment, or water has been assessed by analysis of conserved marker genes, e.g., 16S rRNA genes. In addition, other conserved genes, such as *recA* or *radA* and genes encoding heat shock protein 70, elongation factor Tu, or elongation factor G have been employed as markers for phylogenetic analyses.

The direct analysis of 16S rRNAgene sequences in the environment can be used to study the diversity of microorganisms without culturing. Initially, sequencing of 16S rRNAs were performed using reverse transcriptase. However, formation of secondary structures in rRNA sequences hampers the sequencing of full length

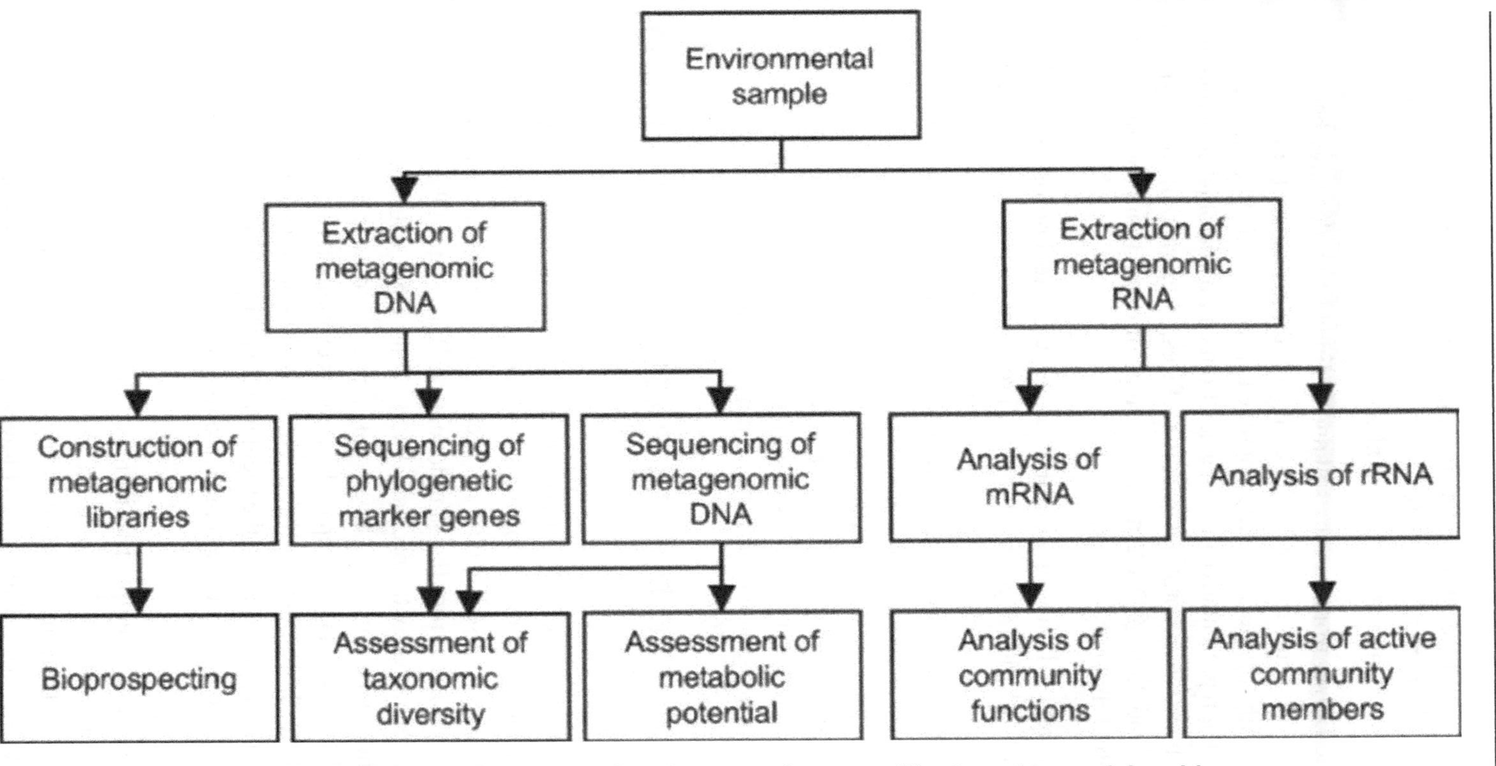

Fig. 9.1. Metagenomic analysis of environmental communities based on nucleic acids

genes. Development of PCR and cyclic sequencing technology resulted in large-scale sequencing of complete 16S rDNA genes, which accelerates the discovery of diverse taxa from various environmental sources.

Metagenomic DNAs from various soil and sediment samples have been used for the estimation of total microbial diversity of the associated environment. However, the validity of metagenomic approach in microbial diversity analysis depends on obtaining representative nucleic acids from entire microbial communities. Incomplete cell lysis,DNA sorption to soil surfaces, coextraction of enzymatic inhibitors and degradation of DNA at various steps of extraction procedures may hinder the microbial diversity analysis.

The employment of next-generation sequencing technologies, such as pyrosequencing of 16S rRNA gene amplicons, provided unprecedented sampling depth compared to traditional approaches, such as denaturing gradient gel electrophoresis (DGGE), terminal restriction fragment length polymorphism (T-RFLP) analysis, or Sanger sequencing of 16S rRNA gene clone libraries. However, the intrinsic error rate of pyrosequencing may result in the overestimation of rare phylotypes. Each pyrosequencing read is treated as a unique identifier of a community member, and correction by assembly and sequencing depth, which is typically applied during genome projects, is not feasible .

9.2. METAGENOMIC DNA EXTRACTION FOR DIVERSITY ANALYSIS

Due to the higher DNA yields, direct lysis methods are reported to access larger fractions of indigenous microbial population. Indirect DNA extraction methods are known to access less diverse population, because these methods relying on the release of microorganisms from the sample matrix. But greater DNA yield is not always positively correlated with species richness. Thus, DNA extracted by indirect methods need not necessarily be less representative of the total microbial community. In few cases, DNA prepared by indirect methods represented more diverse species than those obtained by direct lysis. The amount of co-extracted eukaryotic DNA may also influence the yield and diversity correlation. Therefore, nonbacterial DNA pools should be considered to assess the population dynamics and diversity.

Many workers have attempted to increase DNA yields from soils by using severe physical treatments for longer time. However, treatments can shear DNA, which may not be suitable for diversity analysis based on PCR using Taq DNA polymerase due to the risk of forming chimeric products with smaller template

DNA fragments.Therefore, the metagenomic DNA extraction for microbial diversity analysis should ensure the unbiased lysis of bacterial cells with minimal DNA shearing.

9.3. Direct extraction of rRNA

The use of rRNA, rather than genes coding for rRNA (rDNA), as the target nucleic acid for 16S rRNA probes has been proposed to be advantageous, since multiple ribosomes in bacterial cells may provide an increased sensitivity of detection. However, RNA is highly susceptible to degradation by RNases during extraction procedures and therefore, it is more difficult to recover RNA than DNA from soil. In addition, gel filtration resins such as Sepharose 2B are unsuitable for RNA purification, because low molecular weight RNAs co-elute with humic acids. Due to these complexities, rDNA based phylogeny is widely used for molecular diversity analysis and the direct targeting of 16S rRNA has received comparatively less attention.

Alternatively, direct isolation of ribosomes from soil samples and subsequent extraction yielded rRNA free from contaminants. After the cell lysis, ribosomes are separated by centrifugation steps and rRNA then extracted, which avoids humic acid contamination and RNA degradation

9.4. ESTIMATION OF SOIL MICROORGANISMS

Several strategies have been proposed to estimate the numbers of soil microorganisms. Earlier, substrate-induced respiration and chloroform fumigation extraction have been used to estimate the microorganisms based on microbial respiration. In recent days, the 42 ,6-diamino-2-phenylindole (DAPI) staining method is popular for counting numbers of soil microorganisms. However, due to the complications associated with operation, reproducibility and the high cost, these techniques can not be used in routine microbial ecological studies. Therefore, analysis of metagenomic DNA has been proposed as a strategy for evaluating numbers of soil microorganisms. There exists a linear proportional relationship between soil bacterial biomass and the amount of DNA isolated. Therefore, the bacterial biomass could be evaluated by quantifying levels of environmental DNA. However, coextraction of extracellular DNA should be considered, which may lead to the overestimation of number of living bacteria.

9.5. ESTIMATION OF ACTIVE VERSUS TOTAL BIODIVERSITY

Recovery of intact RNA from environmental samples and the RNA/DNA ratio is an important indicator of the metabolic status of microbial communities.The amount of cellular DNA is relatively stable and can be used to measure the number of cells present; whereas the amount of RNA will change with growth rate and, thus, can be used to measure the metabolic activity. Linear correlation between the RNA/DNA ratio and growth rate were observed in cultured organisms. At slower growth rates, the RNA/DNA ratios become less than 1. Species-specific probes based quantification of 16S rRNA, 16S rDNA and the rRNA:rDNA ratio can be used to measure activity of a particular species in a mixed community.This ratio may address the questions concerning whether the response of a microbial community to environmental change is due to a population increase or activity increase. To obtain a reliable RNA/DNA ratio, both RNA and DNA should be recovered from environmental samples without bias. However, unbiased recovery of both DNA and RNA is a significant challenge due to microbial heterogeneity in natural environments, variations in experimental conditions, and differences in interactions of DNA and RNA molecules with environmental matrices.

Although it is difficult to eliminate all sources of variation, variation originating from microbial heterogeneity and extraction conditions can be minimized if the RNA and DNA are simultaneously extracted from the same fraction of the cell lysate **(Fig. 9.2).** The separation of RNA and DNA using commercial anion exchange resins works better than pH-based differential organic extraction. Similarly, the rapid coextraction of RNA and DNA for the comparison of bacterial diversity by 16S rRNA reverse transcription-PCR (RT-PCR) and 16S rDNA PCR may reveal the active versus total biodiversity of the environment.

Griffiths *et al.* (2000) in an experiment divided the cell extract into two aliquots and processed for the isolation of DNA and RNA simultaneously. Half of the sample was incubated at 37 °C with RNase A to obtain pure DNA and the other half with treated with RNase-free DNase to obtain RNA. The PCR and RT-PCR analysis demonstrated that the difference existed between these profiles, presumably due to the active or total diversity assessed by rRNA or rDNA, respectively.

9.6. NEXT-GENERATION SEQUENCING PLATFORMS

Introduction of next-generation sequencing platforms, such as the Roche 454 sequencer, the SOLiD system of Applied Biosystems , and the Genome Analyzer of Illumina, had a big impact on metagenomic research. The advances in throughput and cost reduction have increased the number and size of metagenomic sequencing projects, such as the *Sorcerer II* Global Ocean Sampling (GOS) project and the metagenomic comparison of 45 distinct microbiomes and 42 viromes. The analysis of the resulting large data sets allowed the exploration of the taxonomic and functional biodiversity and of the system biology of diverse ecosystems.

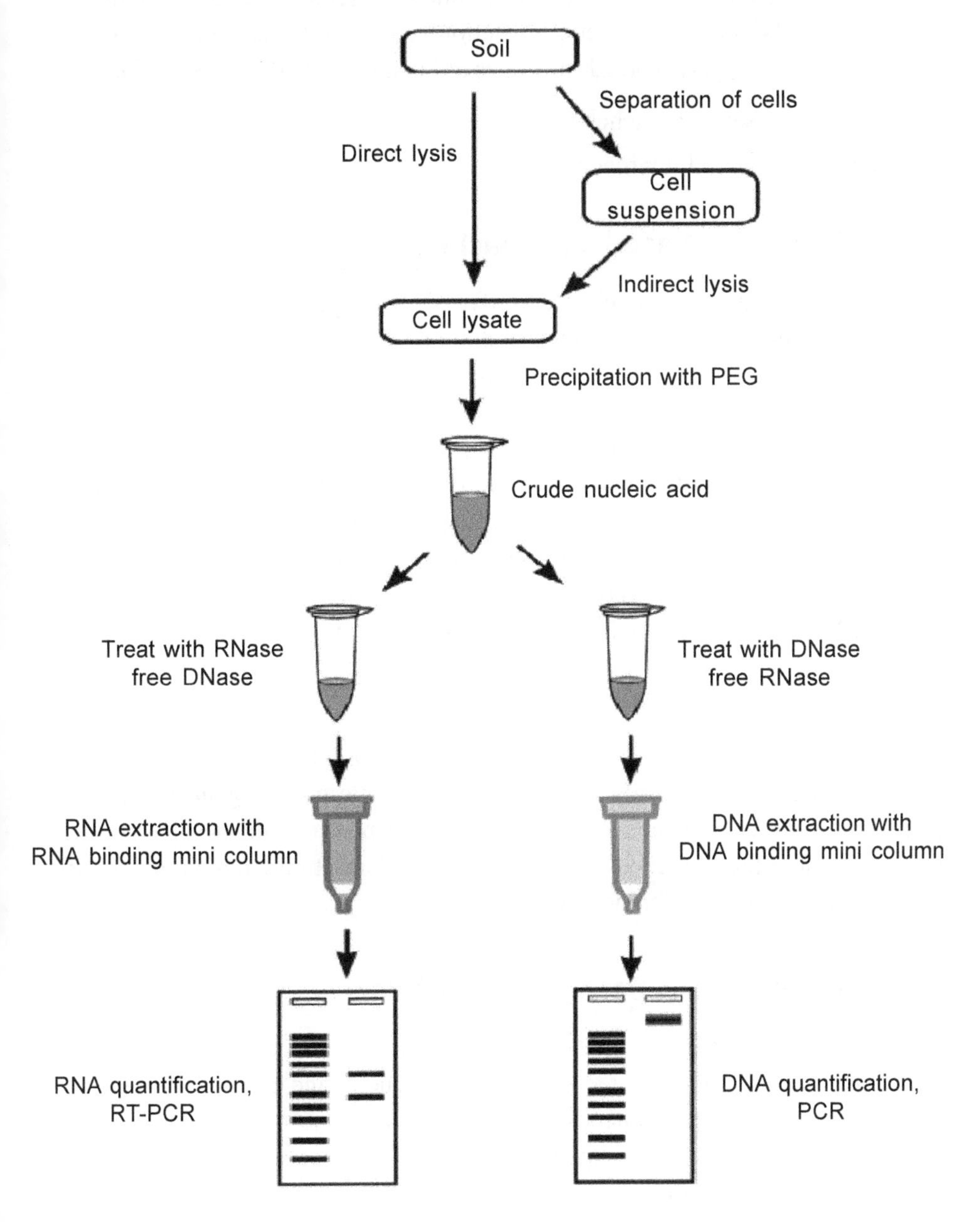

Fig. 9.2. Simultaneous extraction of RNA and DNA

9.7. TAXONOMIC DIVERSITY OF MICROBIAL COMMUNITIES

A crucial step in the taxonomic analysis of large metagenomic data sets is called binning. Within this step, the sequences derived from a mixture of different organisms are assigned to phylogenetic groups according to their taxonomic origins. Depending on the quality of the metagenomic data set and the read length of the DNA fragments, the phylogenetic resolution can range from the kingdom to the genus level. Currently, two broad categories of binning methods can be distinguished:

i) similarity-based approaches
ii) composition-based approaches.

9.7.1. Similarity-based approaches

The similarity-based approaches classify DNA fragments based on sequence homology, which is determined by searching reference databases using tools like the Basic Local Alignment Search Tool (BLAST). Examples of bioinformatic tools employing similarity-based binning are the Metagenome Analyzer (MEGAN), CARMA or the sequence ortholog-based approach for binning and improved taxonomic estimation of metagenomic sequences (Sort-ITEMS). CARMA assigns environmental sequences to taxonomic categories based on similarities to protein families and domains included in the protein family database (Pfam) whereas MEGAN and Sort-ITEMS classify sequences by performing comparisons against the NCBI, nonredundant and NCBI nucleotide databases

One pitfall of these approaches is that taxonomic classification of the metagenomic data sets relies on the use of reference databases that contain sequences of known origin and gene function. To date, the common databases are biased toward model organisms or readily cultivable microorganisms. This is a major limitation for taxonomic classification of microbial communities in ecosystems, as up to 90% of the sequences of a metagenomic dataset may remain unidentified due to the lack of a reference sequence.

9.7.2. Composition-based approaches

Composition-based binning methods analyze intrinsic sequence features, such as GC content codon usage or oligonucleotide frequencies and compare these features with reference genome sequences of known taxonomic origins. Tools such as PhyloPythia, TETRA and the taxonomic composition analysis method (TACOA) allow direct classification of short single reads.

9.8. WEB-BASED METAGENOMIC ANNOTATION PLATFORMS

Recently, Web-based metagenomic annotation platforms, such as the metagenomics RAST (mg-RAST) server, the IMG/M server or JCVI Metagenomics Reports (METAREP) have been designed to analyze metagenomic data sets. Via generic interfaces, the uploaded environmental data sets can be compared to both protein and nucleotide databases, such as the Gene Ontology (GO) database the Clusters of Orthologous Groups (COG) database and the Pfam ,NCBI SEED, and Kyoto Encyclopedia of Genes and Genomes (KEGG) databases.

In this way, multiple metagenomic data sets derived from various environments can be compared at various functional and taxonomic levels. Recent examples of metagenomic surveys of whole microbial communities include those studying the hindgut microbiota of a wood-feeding higher termite ,glacier ice sludge communities subjected to enhanced biological phosphorus removal a biogas plant microbial community, and Minnesota farm soil.

9.9. TRACKING OF RECOMBINANT MICROORGANISMS

The possibility of genetically engineered microorganisms to be used in environmental applications requires an understanding of the fate of recombinant DNA introduced into the environment. To study the fate of a genetically engineered organism introduced into a soil environment, direct extraction of DNA followed by a specific quantitative detection based on PCR or hybridization are used to determine the persistence of the recombinant gene

In an experiment by Tien *et al.* (1999) found that metagenomic DNA prepared from soil samples by direct lysis contained fragments of various molecular sizes. However, when *E. coli* was added to acidwashed sea-sand and subjected to the same extraction procedure, no such fragments were detected. Therefore, it was argued that the small sized DNA fragments could be the extracellular DNA at various stages of degradation.

9.9.1. Horizontal gene transfer in soil

In several investigations, higher yields of DNA from soils or sediments are usually obtained with the direct lysis method.The higher yields by direct method could be at least partly because it extracts the extracellular DNA also. DNA released from organisms constitutes the extracellular fraction, which is subjected to the degradation by soil DNases. However, extracellular DNA that tightly adsorbed to soil particles is protected from DNase activity in the soil. The extracellular DNA fragments, which are partially degraded by DNases involved in horizontal gene transfer. Thus, persistence of a gene does not necessarily correlate with the survival rate of an introduced organism, because recombinant DNA can also persists extracellularly or may be transferred into indigenous microorganisms.

Some genes transferred are maintained in the environment by native microbial cells. For instance, it has been demonstrated that a plasmid, pJP4, containing genes coding for metal resistance, transferred from *E. coil* to indigenous soil populations and stably maintained in the environment by native microorganisms. These factors should be considered while tracking recombinant microorganisms in the environment.

9.10. SINGLE CELL GENOMICS

In spite of the existence of millions of different genomes in the environment, elucidation and assembly of complete genome of an uncultured organism has become possible. Eliminating the community complexity by initial selection and picking the single cell is used to construct genomic library of specific microorganism. In this approach, intact cells are isolated from the environmental sample. To detect and collect the single cells of interest the following methods are proposed.

a) Fluorescence activated cell sorting (FACS)
b) Micromanipulation
c) Microfluidics

9.10.1. Whole genome amplification (WGA) methods for single cell isolation

Isolation of single cells alone has not solved the problem of uncultured organism, because single cells do not provide sufficient DNA for genomic sequencing. Development of whole genome amplification (WGA) methods has made the sequencing of genome from a single cell possible.

9.10.1.1 *Multiple displacement amplification (MDA)*

A method called multiple displacement amplification (MDA) can amplify the few femtograms of DNA in a bacterial cell up to micrograms. In MDA reaction, random hexamers and the DNA polymerase from bacteriophage phi29 are used to amplify DNA in a 30 °C isothermal reaction. The Phi29 polymerase has an efficient strand displacement activity, which makes multiple copies from each template as the polymerase synthesizes new strands while concurrently displacing previously extended strands (Fig 9.3). In addition, the Phi29 polymerase is a highly processive enzyme with an efficient proofreading capacity, which ensures the fidelity of the genome sequences. It has been successfully demonstrated that the genome of a single *E. coli* cell or a single human sperm could be amplified by MDA. The MDA-based single cell genome amplification and subsequent shotgun analysis can be used to assemble the complete genome. The MDA amplicons average 12 kb in length and range up to >100 kb making them suitable for DNA library construction and Sanger sequencing or pyrosequencing.

9.10.1.2. *Limitations of multiple displacement amplification (MDA)*

MDA may not yield uniform and unbiased amplification of the entire genome from the single copy present in most bacteria. Amplification bias in the MDA reaction can result in under representation and loss of some sequences.

9.10.1.3. *MDA on Pooled single cells*

Several studies have reported only about 50–75% completion of genomes from uncultured single cells including a novel soil bacteria a marine organism *Prochlorococcus* TM7 from soil and human oral cavity, a candidate phylum for which no sequenced members had existed. To obtain the complete genome sequence, MDA amplicons derived from several single cells of the same strain can be pooled and subsequently used for shotgun analysis. Alternatively, single cells may be pooled before the MDA reaction. Recently, the complete genome of an uncultured bacteria derived from protists inhabiting the termite gut has been reported. This genome was reconstructed by carrying out MDA on pooled single cells derived from a single protist host cell. Furthermore, partial genome sequences obtained from single cells may be combined with whole metagenome shotgun sequences to map the complete genome.

9.10.1.4. Advantages of Whole genome amplification through MDA

Whole genome amplification through MDA is not only useful for single cell genomics; but also dramatically increases the possibilities of cloning and sequencing of metagenomic DNA. Environments with relatively high biomass, such as agricultural soils and forest soils, yield sufficient quantity of DNA for metagenomic applications. But, DNA from soil samples with very low cell counts, such as contaminated subsurface sediment cores, is not readily accessible.

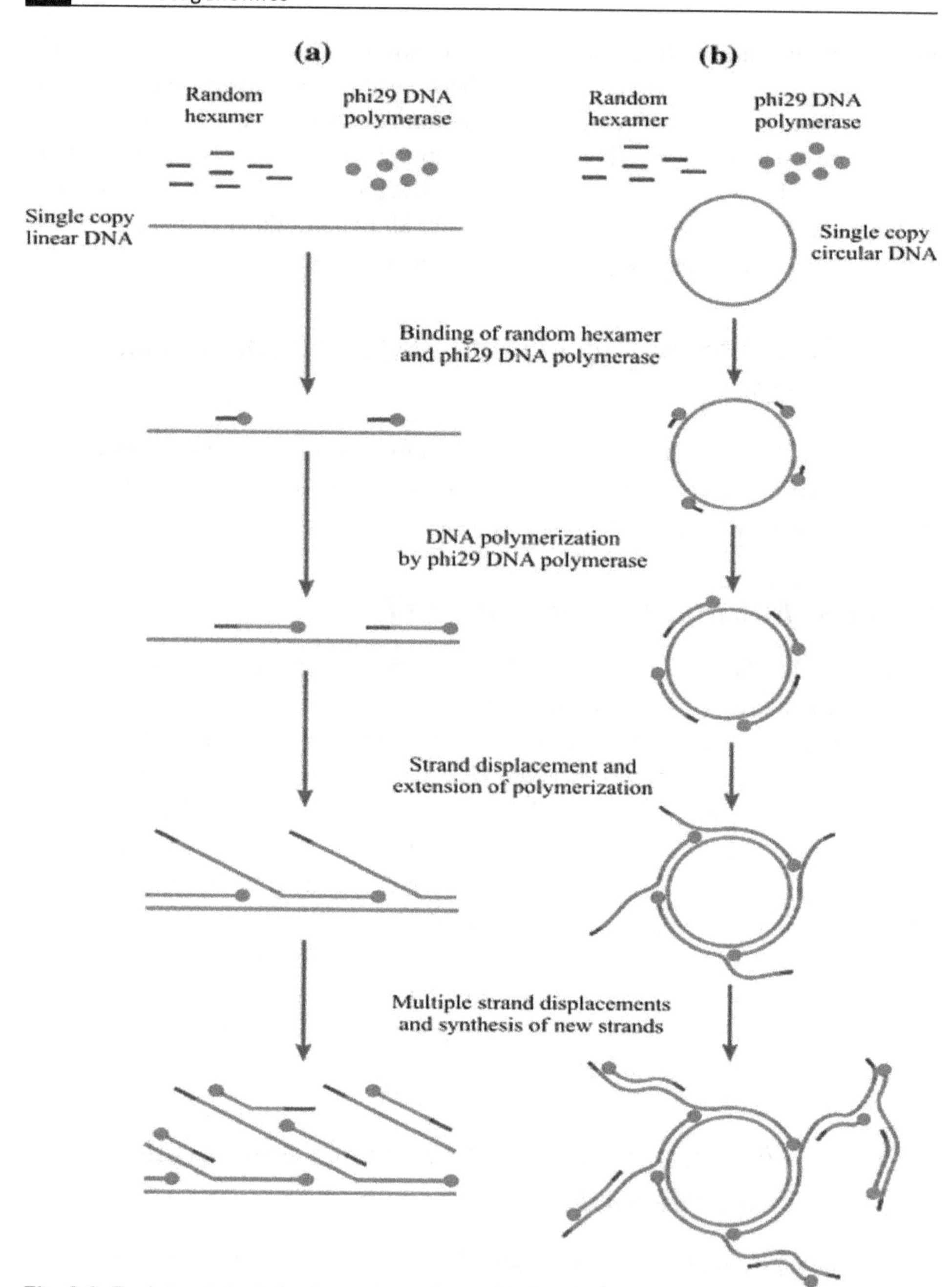

Fig. 9.3. Enrichment of single genome by multiple displacement amplification (MDA). The random hexamers bind to the denatured DNA and the phi29 DNA polymerase extends the strand until it reaches newly synthesized double-stranded DNA. The enzyme proceeds polymerization by displacing the double strand. The random primers bind to the newly synthesized single-stranded region and polymerization starts on the new strands, resulting in whole genome amplification of the DNA of single cells. (a) MDA of a linear DNA fragment; (b) MDA of a circular single stranded DNA.

The MDA has also been considered as a pre-PCR to access the low copy number target sequences from the environment. Huge amount of sample would be needed to isolate sufficient DNA for traditional library construction. If larger quantities of sample are used, the level of contaminants also proportionately increased, which affects the cloneability of the isolated metagenomic DNA.

MDA followed by standard PCR results in successful PCR amplification, whereas direct PCR could not amplify the lowcopy number target sequences from contaminated sediments.

In addition, it has been demonstrated that phi29 DNA polymerase is more tolerant to inhibition by humic substances than Taq DNA polymerase.

MDA based whole-community genome amplification before microarray hybridization and reported that this approach is capable of identifying subnanogram quantities of microbial DNA. Because microarray hybridization alone are not sufficient for the identification of less dominant microbial populations, because of their poor representation in the total metagenomic DNA

Overall, the MDA reduces amount of soil sample and it yields large quantities of pure cloneable DNA for various applications.

9.11. METATRANSCRIPTOMICS

Metagenomics provides information on the metabolic and functional capacity of a microbial community. However, as metagenomic DNA-based analyses cannot differentiate between expressed and non expressed genes, it fails to reflect the actual metabolic activity. Recently, sequencing and characterization of metatranscriptomes have been employed to identify RNA-based regulation and expressed biological signatures in complex ecosystems (Fig 9.1).

So far, metatranscriptomic studies of microbial assemblages *in situ* are rare. This is due to difficulties associated with the processing of environmental RNA samples. Technological challenges include a) the recovery of high-quality mRNA from environmental samples b) short half-lives of mRNA species (c) separation of mRNA from other RNA species.

Until recently, metatranscriptomics had been limited to the microarray/high-density array technology or analysis of mRNA-derived cDNA clone libraries. These approaches have produced significant insights into the gene expression of microbial communities but have limitations. A microarray gives information about only those sequences for which it was designed. The detection sensitivities are not equal for all imprinted sequences, as results are dependent on the chosen hybridization conditions. Low-abundance transcripts are often not detected. Although transcript cloning avoids some of these problems through random amplification and sequestering of mRNA fragments, it introduces other biases associated with the cloning system and the host of the libraries.

The limitations of both approaches can be circumvented by application of direct cDNA sequencing employing next-generation sequencing technologies. This provides affordable access to the metatranscriptome and allows whole-genome expression profiling of a microbial community. In addition, direct quantification of the transcripts is feasible. Leininger and his co workers (2006) first employed pyrosequencing to unravel active genes of soil microbial communities. In this way, the activity and importance of ammonia-oxidizing archaea in soil ecosystems have been shown.

Other metatranscriptomic studies employing direct sequencing of cDNA have targeted the ocean surface waters from the North Pacific subtropical gyre, coastal waters to Bergen, Norway, a phytoplankton bloom in the Western English Channel, and soil samples from a sandy lawn.

Small RNAs (sRNAs) involves in many environmental processes, such as carbon metabolism and nutrient acquisition. cDNA sequences are not homologous to known genes encoding proteins. Almost a third of these unassigned cDNA sequences show similarities to intergenic regions of microbial genomes in which sRNA molecules are encoded. Thirteen known sRNA families were identified in the metatranscriptomic data set by searching the RNA family database Rfam. These sequences display characteristic conserved secondary structures and were often flanked by potential regulatory elements. This indicates the presence of so-far-unrecognized putative sRNA molecules and provides evidence for the importance of sRNAs for the regulation of microbial gene expression in response to changing environmental parameters.

9.12. METAPROTEOMICS

The proteomic analysis of mixed microbial communities is a new emerging research area which aims at assessing the immediate catalytic potential of a microbial community. In 2004, Wilmes and Bond coined the term "metaproteomics" as a synonym for large-scale characterization of the entire protein complement of environmental microbiota at a given time point.

In this landmark study, the proteins produced by a microbial community derived from activated sludge were analyzed by two-dimensional polyacrylamide gel electrophoresis and mass spectrometry. Highly expressed proteins, such as an outer membrane protein and an acetyl coenzyme A acyltransferase, were identified. These enzymes putatively originated from an uncultured polyphosphate-accumulating *Rhodocyclus* strain that was dominant in the activated sludge.

9.12.1. Examples for metaproteomics studies

One of the most comprehensive metaproteomic studies has been conducted in natural acid mine drainage microbial biofilm by mass spectrometry to study the gene expression, key activities, and metabolic functions. In this way, more than 2,000 proteins from the five most abundant microorganisms were identified. In addition, 357 unique and 215 novel proteins were detected. One highly expressed novel protein was capable of iron oxidation, a process central to acid mine drainage formation. This study and other studies provided comprehensive insights into microbial communities that exhibited a relatively low complexity *i.e.*, communities derived from a continuous-flow bioreactor fed with cadmium ,activated sludge and the phyllosphere.

In addition, metaproteomic analyses of microbial communities displaying a high complexity,such as communities present in the hindguts of termites, sheep rumens, human fecal samples, human saliva samples, marine samples, dissolved organic matter from lake and forest soil and contaminated soil and groundwater, were carried out but at a lower resolution. Nevertheless, it is a daunting task to detect and identify all proteins produced by a complex environmental microbial community.

9.12.2. Challenges in metaproteomics

i) uneven species distribution,

ii) broad range of protein expression levels within microorganisms,

iii) large genetic heterogeneity within microbial communities

Despite these hurdles, metaproteomics has a huge potential to link the genetic diversity and activities of microbial communities with their impact on ecosystem function.

Mini Quiz

1. How can you differentiate active diversity from total microbial diversity?
2. What are the key differences between similarity and sequence based approaches?
3. What is RAST?
4. Explain single cell genomics and the methods involved?
5. Elucidate the merits and demerits of multiple displacement amplification in single cell isolation?

CHAPTER - 10

METAGENOMICS AND ECOLOGY

The exigent questions in microbial ecology focus on how microorganisms form symbioses with eukaryotes, compete and communicate with other microorganisms, and acquire nutrients and produce energy. Thus far, metagenomics has provided insights into each of these areas, but in each instance, the challenge is to link the genomic information with the organism or ecosystem from which the DNA was isolated. Expression of a gene in a cultured host can establish gene function, but without the appropriate biological context, circumspection is required in drawing ecological inferences. Future technical innovations are needed to extend insights from metagenomics from inference to mechanistic analyses.

10.1. SYMBIOSIS

Many bacterial symbionts that have highly specialized and ancient relationships with their hosts do not grow readily in culture. Many of them live in specialized structures, often in pure or highly enriched culture, in host tissues, making them ideal candidates for metagenomic analysis because the bacteria can be separated readily from host tissue and other microorganisms.This type of analysis has been conducted with *Cenarchaeum symbiosum*, a symbiont of a marine sponge, a *Pseudomonas*-like bacterium that is a symbiont of *Paederus* beetles, *Buchnera aphidicola*, an obligate symbiont of aphids, the Actinobacterium *Tropheryma whipplei*, the causal agent of the rare chronic infection of the intestinal wall and the Proteobacterium symbiont of the deep sea tube worm *Riftia pachyptila.*

These systems provide good models for metagenomic analysis of more complex communities and thus warrant further attention in this review, although the term metagenomics typically connotes the study of multispecies communities. Therefore, the following section focuses on two of these obligate symbionts and the insight into their lifestyles offered by metagenomic analysis.

10.1.1. *Buchnera*-aphid symbiosis

The first genome reconstruction of an uncultured organism was that of *Buchnera aphidicola*, the endosymbiont of aphids. The relationship between the bacterium and the insect is ancient, leaving each partner unable to function independently of the other, as is reflected in the genomic analysis. Moran's group isolated bacterial DNA from the insect and sequenced and reassembled the bacterial genome. The genus *Buchnera* contains a "reduced" genome of 564 open reading frames. Upon comparison with a reconstructed ancestral genome, 1,906 genes appear to have been lost. Most of the functions are associated with biosynthetic pathways contributed by the host, suggesting that the genome shrinkage is the result of the symbiotic lifestyle, which has become obligate because of gene loss. The reconstruction of *B. aphidicola*'s genome provided insights into the evolution of the symbiosis between the insect and bacterium, the biochemical mutual dependence that they have developed, and the mechanisms of genome shrinkage and rearrangement. The success of genome reconstruction with a single uncultured species provided part of the impetus needed to propose sequencing and reconstructing genomes in more complex assemblages.

10.1.2. Proteobacterium-tube worm symbiosis

Riftia pachyptila, the deep sea tube worm, lives 2,600 m below the ocean surface, near the thermal vents that are rich in sulfide and reach temperatures near 400°C. The tube worm does not have a mouth or digestive tract, and therefore it is entirely dependent on its symbiotic bacteria, which provide the worm with food. The bacteria live in the trophosome, a specialized feeding sac inside the worm. The bacteria and trophosome constitute more than half of the animal's body mass. The bacteria oxidize hydrogen sulfide, thereby producing the energy required to fix carbon from CO_2, providing sugars and amino acids (predominantly as glutamate) that nourish the worm. The worm contributes to the symbiosis by collecting hydrogen sulfide, oxygen, and carbon dioxide and transporting them to the bacteria on hemoglobin-like molecules.

The bacterium is a member of the *Proteobacteria*, as identified by 16S rRNA gene sequence. The bacteria have not been grown in pure culture in laboratory media, but they provide an excellent substrate for metagenomics because they reach high population density in the trophosome and exist there as a single species. A gene similar to ribulose-1,5-bisphosphate carboxylase/oxygenase (RubisCO) was identified from the fosmid library. All of the residues associated with the active site are conserved in the protein sequence deduced from the DNA sequence, and it has highest similarity with the RubisCO from *Rhodospirillum rubrum*. The characterization of this gene lends further support to the premise that the chemoautotrophic bacterial symbiont in *R. pachyptila* fixes carbon for its host. The libraries were also screened for two-component regulators with a labeled histidine kinase gene as a probe. They identified a two-component system whose components complemented an *envZ* and a *phoR creC* double mutant, respectively.The discovery of a functional *envZ* homologue indicates that the symbiont carries a response regulator that is typical of *Proteobacteria*, although the signals eliciting responses from these proteins have not yet been identified.

Genomic analysis of the symbiont also led to the identification of a gene encoding flagellin, which was expressed in *E. coli* and shown to direct the synthesis of flagella that are immunologically cross-reactive with *Salmonella* flagella. The presence of genes for flagella suggested to the authors that the endosymbiont has a free-living stage in its life cycle and may infect each generation of tube worms rather than being passed maternally.

10.2. COMPETITION AND COMMUNICATION

Competition for resources among community members selects for diverse survival mechanisms, including antagonism and mutualism among the members. Understanding these mechanisms is central to advancing the definition of principles that govern microbial community structure, function, and robustness. Historically, genetics has provided the most convincing evidence for traits contributing to microbial fitness. Classic mutant analysis has revealed genes required for microbial competition, antagonism, and mutualism. Mutant analyses have provided the greatest advances in knowledge because screening mutants containing random mutations for effects on fitness has led to the identification of genes that would not have been predicted to play a role in microbial competition or mutualism.

Genes for competition and cooperation are hard to recognize based on sequence alone because the utility of their functions is entirely dependent on ecosystem context and the nature of the resources that are limiting. Therefore, genomics by itself does not provide a means to test ecological hypotheses or identify genes that confer fitness, but it can provide the basis for forming hypotheses. Ecological hypotheses are difficult to test in microorganisms that cannot be cultured or for which there are no genetic tools; however, functional genomics coupled with chemical ecology can yield informative answers.

Chemical ecology involves the identification of small molecules with biological activity and proposed ecological function. These compounds can be identified through a variety of methods, including metagenomics. The addition of these molecules to communities can provide the basis for postulating their ecological roles in the community by measuring perturbations of community function. The following sections explore the discovery of small molecules in metagenomic libraries and postulate the ecological functions of these molecules in the organisms producing them.

10.3. ROLE OF SMALL MOLECULES

Small-molecule discovery by functional metagenomics has concentrated on antibiotics, which are of interest for their pharmaceutical applications as well as for their roles in ecosystem function. Traditional antibiotic screens for molecules that inhibit bacterial growth have led to the discovery of antibiotics in metagenomic libraries. They have not been a rich source of novel antibiotics, likely because of the experimental limitations associated with the search. In studies that report frequencies, antibiotic-producing clones are detected at a frequency of approximately 1 producer per 10^4 clones. This low frequency hinders discovery because space and labor are required to conduct typical antimicrobial screens.

With standard inhibition assays, a *Mycobacterium*-inhibiting antibiotic, terragine, was discovered from a soil metagenomic clone maintained in *Streptomyces lividans* and acyltyrosines from a clone maintained in *E. coli*. Colored antibiotics represent a disproportionate share of those discovered because they can be identified visually. For instance, a clone noticed for its brown pigment was found to produce melanin, which masked orange and red pigments, two novel antibiotics, turbomycin A and turbomycin B. A purple pigmented clone produced violacein, previously shown to be an antibiotic made by the soil bacterium *Chromobacterium violaceum*. The sequence of the genes on the metagenomic clone diverged substantially from the *C. violaceum* violacein biosynthetic operon despite similar genetic organization suggesting that the pathway on the metagenomic clone was derived from an organism other than *C. violaceum*. Osburne's group identified structurally related compounds, indirubin and indigo blue, in a soil metagenomic DNA library based on their blue color.

10.3.1. Sequence-based screening for small molecules

The first polyketide synthases, enzymes involved in synthesis of polyketides, the broad class of antibiotics that includes erythromycin, epithilone, and rifamycin, were first cloned from soil with a PCR based approach.This approach was adapted for screening metagenomic libraries by Osburne's group, who screened a 5,000-member metagenomic library for conserved regions of genes encoding type I polyketide synthase. Primers directed toward a conserved region of polyketide synthase I genes that flanks the active site of the ketoacyl synthetase domain were used to screen pools of clones. In addition, screening clones in both *E. coli* and *Streptomyces lividans* by chemical means revealed two novel compounds, fatty dienic alcohol isomers.

10.3.2. Antibiotics as signal molecules

If antibiotics evolved as mediators of functions other than warfare such as communication, antibiotic discovery will be expedited by screening metagenomic clones for signaling compounds as well as inhibitory compounds. The challenge is to develop assays that detect signaling by many compounds. A surprising result from the Davies group indicated that subinhibitory concentrations of many antibiotics induce quorum sensing despite no resemblance in structure to the acylated homoserine lactones that appear to be the natural inducers. This result presents a propitious opportunity—a single screen might capture molecules that are quorum-sensing inducers as well as antibiotics. This opportunity was investigated by designing a high throughput screen to identify compounds that induce the expression of genes under the control of a quorum-sensing promoter. The screen is intracellular, meaning the metagenomic DNA is in the same cell as the sensor for quorum-sensing induction. The sensor is comprised of the *luxR* promoter, which is induced by acylated homoserine lactones, linked to *gfp*, and resides on a plasmid in an *E. coli* strain that did not induce quorum sensing itself. If an inducer of the *luxR*-mediated transcription of *gfp* is expressed from the metagenomic DNA, the cell fluoresces and can be captured by fluorescence-activated cell sorting or as a colony observed by fluorescence microscopy. Conversely, this sensor system can detect inhibitors of quorum sensing if acylated homoserine lactone is added to the medium and fluorescence-activated cell sorting is set to collect the nonfluorescent cells (Fig. 10.1).

Metagenomic libraries from microbiota of the soil and from the midgut of the gypsy moth have been subjected to this screen, and an array of genes has been identified. Their products are under analysis, and some appear to differ from previously described quorum sensing inducers.

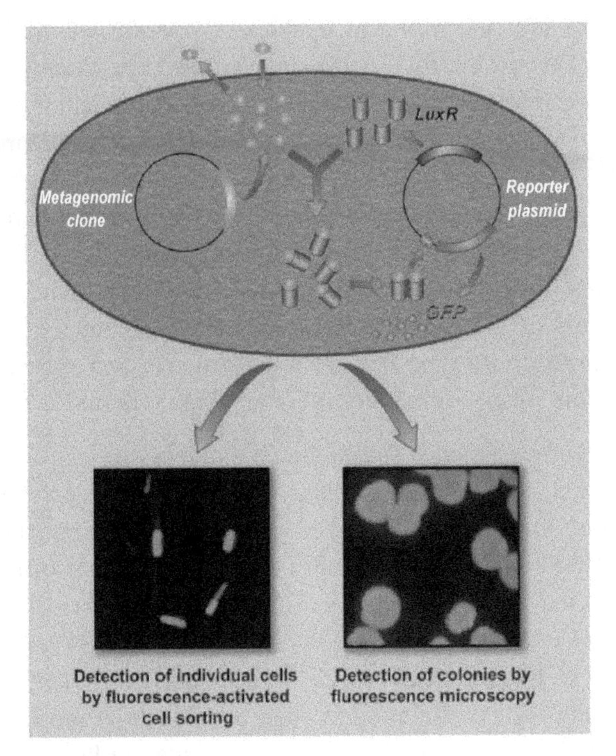

Fig. 10.1. Intracellular screen for quorum-sensing inducers

10.4. BIOGEOCHEMICAL CYCLES

10.4.1. Acid mine drainage

An exciting potential of metagenomics is to provide community-wide assessment of metabolic and biogeochemical function. Analysis of specific functions across all members of a community can generate integrated models about how organisms share the workload of maintaining the nutrient and energy budgets of the community. The models can then be tested with genetic and chemical approaches. The best example of such an analysis is the nearly complete sequencing of the metagenome of a community in acid drainage of the Richmond mine, which represents one of the most extreme environments on Earth.

The microbial community forms a pink biofilm that floats on the surface of the mine water. The drainage water below the biofilm has a pH of between 0 and 1 and high levels of Fe, Zn, Cu, and As (317, 14.4, and 2 mM, respectively). The solution around the biofilm water is 42°C and microaerophilic. There is no source of carbon or nitrogen other than the gaseous forms in the air. The community is dominated by a few bacterial genera, *Leptospirillum*, *Sulfobacillus*, and sometimes

Acidomicrobium, and one archaeal species, *Ferroplasma acidarmanus*, and other members of its group, the *Thermoplasmatales*. The mine is rich in sulfide minerals, including pyrite (FeS_2), which is dissolved as a result of oxidation, which is catalyzed by microbial activity. The simple community structure made it possible to clone total DNA and sequence most of the community with high coverage. The G+C content of each clone provided a good indicator of its source because the G+C content of the genomes of the dominant taxa in the mine differ substantially. Sequence alignment of 16S rRNA and tRNA synthetase genes confirmed the organismal origins of the clones. Nearly complete genomes of *Leptospirillum* group II and *Ferroplasma* type II were reconstructed, and substantial sequence information for the other community members was reported.

The metagenomic sequence substantiated a number of significant hypotheses (Fig. 10.2). First, it appears that *Leptospirillum* group III contains genes with similarity to those known to be involved in nitrogen fixation, suggesting that it provides the community with fixed nitrogen. This was a surprise because the previous supposition was that a numerically dominant member of the community, such as *Leptospirillum* group II, would be responsible for nitrogen fixation. However, no genes for nitrogen fixation were found in the *Leptospirillum* group II genome, leading the investigators to suggest that the group III organism is a keystone species that has a low numerical representation but provides a service that is essential to community function. *Ferroplasma* type I and II genomes contain no genes associated with nitrogen fixation but contain many transporters that indicate that they likely import amino acids and other nitrogenous compounds from the environment. Energy appears to be generated from iron oxidation by both *Ferroplasma* and *Leptospirillum* spp. The genomes of both groups contain electron transport chains, but they differ significantly. The genomes of *Leptospirillum* group II and III contain putative cytochromes that typically have a high affinity for oxygen. The cytochromes may play a role in energy transduction as well as in maintaining low oxygen tension, thereby protecting the oxygen-sensitive nitrogenase complex. All of the genomes in the acid mine drainage are rich in genes associated with removing potentially toxic elements from the cell. Proton efflux systems are likely responsible for maintaining the nearly neutral intracellular pH, and metal resistance determinants pump metals out of the cells, maintaining nontoxic levels in the interior of the cells.

The acid mine drainage community provides a model for the analysis of other communities. Determining the origin of DNA fragments and assigning functions may be more difficult for communities that are phylogenetically or physiologically more complex, but the approach will be useful for all communities.

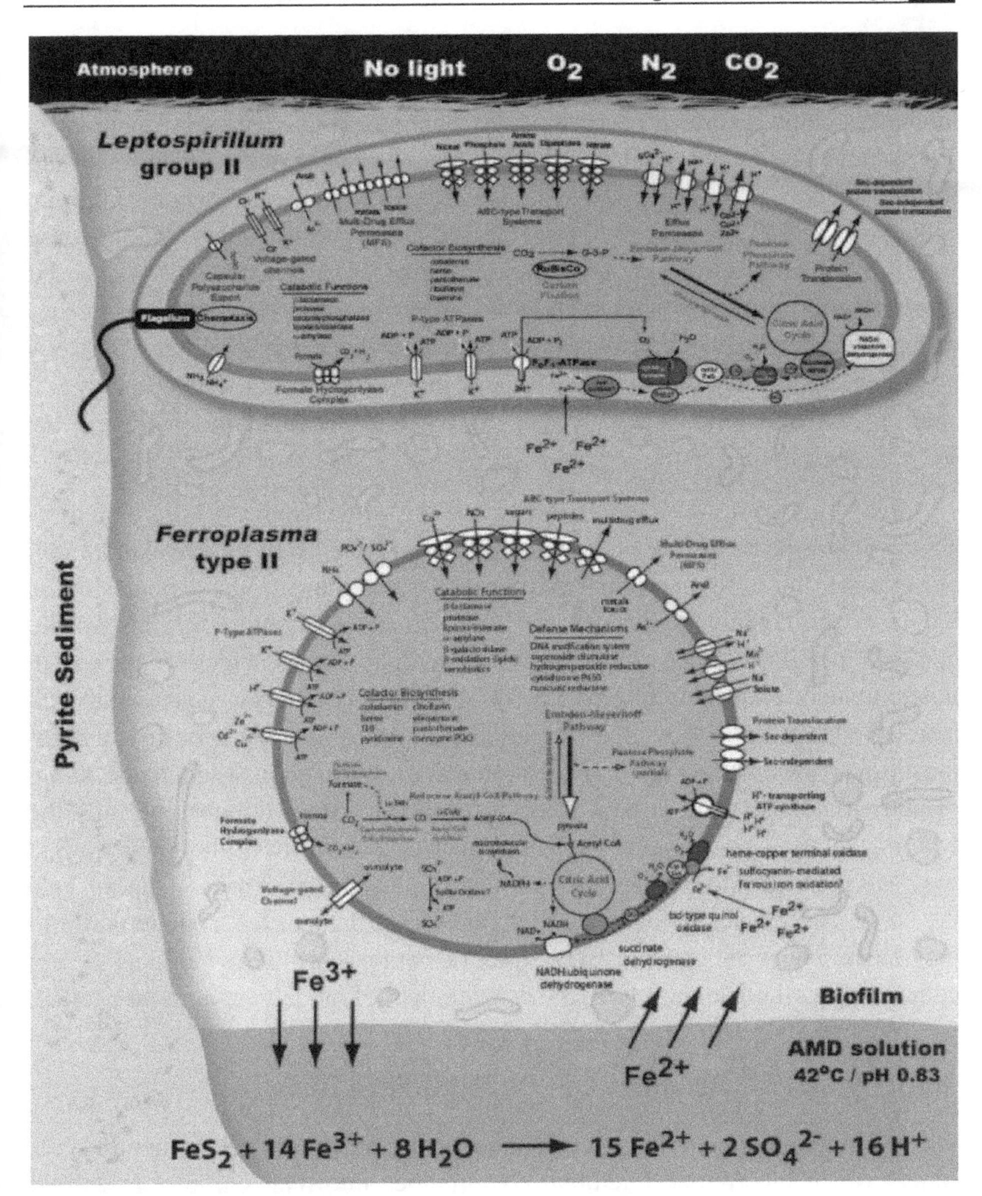

Fig. 10.2. Metagenomics-based model of biogeochemical cycles mediated by prokaryotes in acid mine drainage.

10.4.2. Sargasso Sea

The Sargasso Sea is a complex and physically sprawling ecosystem compared with the contained acid mine drainage system. The inputs and outputs are more difficult to quantify, and the phylogeny of the community members has not been exhaustively surveyed. Venter *et al.* (2004) embarked on the largest metagenomics project to date (affectionately dubbed megagenomics), in which they sequenced over 1 billion bp and claim to have discovered 1.2 million new genes. Intriguing inferences could be drawn because of the sheer size of the data set. They placed 794,061 genes in a conserved hypothetical protein group, which contains genes to which functions could not be confidently assigned. The next most abundant group contained 69,718 genes apparently involved in energy transduction. Among these were 782 rhodopsin-like photoreceptors, increasing the number of sequenced proteorhodopsin genes by almost 10-fold. Linkage of the rhodopsin genes to genes that provide phylogenetic affiliations, such as genes encoding subunits of RNA polymerase, indicated that the proteorhodopsins were distributed among taxa that were not previously known to contain light-harvesting functions, including the *Bacteroides* phylum.

The Sargasso Sea data set is a gold mine for further analysis. Intriguing hints about many aspects of ecosystem function abound and await further exploration. For example, an intriguing initial observation is that many of the genomes in the Sargasso Sea contain genes with similarity to those involved in phosphonate uptake or utilization of polyphosphates and pyrophosphates, which are present in this extremely phosphate limited ecosystem. The phosphorus cycle is not well understood, and this collection of genomes provides a new route for discovery of the mechanisms of phosphorus acquisition and transformation. The understanding of nutrient cycling will be advanced by reconstruction of the genomes and the type of function-species analysis. The sequence data set from the Sargasso Sea provided the means to reassemble a number of genomes with criteria that include the depth of sequencing coverage, oligonucleotide frequencies, and similarity to previously sequenced genomes. The structures of these genomes individually and collectively will no doubt inform the development of models for nutrient cycling.

10.5. METAGENOMICS OF BIOSILCIFICATION

The precipitation of silica in geothermal hot springs leads to the formation of silica sinters which incorporate microbial communities into their edifices. Fossilised microorganisms in active geothermal systems provide a modern analogue to fossilisation/preservation in ancient silica sinters. Silica precipitation in geothermal systems is governed by purely abiotic processes linked to changes in silica saturation, which in turn are linked to changes in temperature, pH and silica concentration upon the emergence of the geothermal waters at the Earth's surface. Microbes are ubiquitous in geothermal systems but what controls their phylogenetic abundance as well as the links between community diversity and geochemical regimes are still poorly understood.

10.6. POPULATION GENETICS AND MICROHETEROGENEITY

Metagenomic analysis has revealed that even apparently uniform populations contain substantial microheterogeneity. Within the population of *Cenarchaeum symbiosum* associated with the marine sponge, the rRNA genes are highly conserved, showing >99.2% identity, which indicates that the population comprises a single species. In the genomic regions flanking the rRNA genes, however, the DNA sequence identity drops to 87.8%. The *Ferroplasma* type II group appears to contain a composite genome, with segments derived from three sources. In contrast,the *Leptospirillum* group II genome contained very few single-nucleotide or large-scale genome polymorphisms. These studies point to the importance of conducting genomic analysis on mixtures of strains to obtain a portrait of the heterogeneity within the species. In fact, metagenomics may provide insight into genome variation of organisms that can be readily cultured. If genetic variation in the environmental population is of interest, it may be more productive to clone the genome from the natural population than analyze the genomes of individuals cultured from it.

10.7. KEY FACTS TO REMEMBER

1. Prokaryotic genomes are highly variable in genome size and gene content among strains from both within and between species.
2. Microbial species with narrow ecological niches generally have smaller genomes than those with broader ecological niches.
3. A large fraction (20–40%) of identified open reading frames in sequenced microbial genomes code for proteins with unknown functions. Most of these genes are likely regulated by ecological-niche specific factors.

Mini Quiz

1. What is the significance of Buchnera-aphid symbiosis as revealed by metagenomics?
2. How functional metagenomics aids in discovering small molecules?
3. What does the ecology of Sargasso Sea reflects?
4. What is the role of *Leptospirillum* and *Ferroplasma* in acid mine drainage?

CHAPTER - 11

MOBILE METAGENOMICS

An exciting extension of the metagenomics is the high-throughput analysis of the "mobilome" or mobile metagenome, the genomics of mobile elements from uncultured organisms. Genes present on mobile genetic elements (MGE) that populate soil ecosystems constitutes the mobilome of the environment.

MGE include plasmids, transposons, insertion sequences and integrons, which may move between bacterial cells in a population or mobilize into a new host species and introduce new genetic material. As the standard metagenomic analysis, mobile metagenomics promise access to all MGE infecting community members regardless of the cultivability of host bacteria. MGEs promote gene flow, or horizontal gene transfer, between species and are a driving force in bacterial adaptation and evolution. However, the mobile metagenome of many ecosystems have not been efficiently explored.

11.1. PLASMID METAGENOME

Autonomously replicating plasmids among the uncultured organisms plays a major role in the metabolic activities in any environment. Several endogenous and exogenous strategies exist for the isolation of plasmids from communities. However, the limitation of both endogenous and exogenous techniques is the requirement of a replication origin and a selectable phenotype marker on the heterologous host. In general, antibiotic or heavy metal resistance is used to identify recipient cells harboring plasmids and thus, plasmids not encoding these genes or unable to express them in the heterologous host cannot be captured. Recently, treatment with a novel exonuclease, "plasmid-safe DNase", is widely used to selectively remove linear DNA fragments. Plasmid-safe DNase selectively removes double-stranded and single-stranded linear DNA, as well as single-stranded circular DNA. However, it does not affect closed circular supercoiled or nicked circular double-stranded DNA molecules, thus effectively enriching the community plasmid DNA. Shot-gun library may be created with enriched plasmid DNA and sequenced to explore the plasmid encoded genes (Fig.11.1).

11.1.2. The transposon-aided capture (TRACA) system

It is a recent strategy developed for accessing the plasmids from environment. In TRACA, plasmids are acquired from metagenomic DNA extracts independent of plasmid encoded traits such as selectable markers or replication origin. Instead, all plasmids in the enriched population are provided with an origin of replication and a selectable marker, by tagging plasmids with a transposon encoding these functions in an in vitro reaction.

For example, using transposon carrying origin of replication from pBR322 and an antibiotic selection marker, plasmids can be captured in *E. coli*. Use of transposons with a greater range of replication origins, selectable markers and a number of host organisms such as *Bacillus, Pseudomonas* and *Streptomyces* may improve the efficiency of TRACA for the isolation of plasmids of various groups of microorganisms. The use of a Gram-positive host and replication origin capture plasmids unstable in classic Gram-negative hosts such as *E. coli*, and facilitate the functional characterization of plasmids acquired from ecosystems predominated by Gram-positive bacteria.

11.2. METAGENOME OF TRANSPOSABLE ELEMENTS

Transposable elements (TE), integrons and genomic islands are also considered as MGE, though they integrate into the bacterial host chromosome, plasmids or bacteriophage. Due to the integrative nature of these elements, they can readily be detected in sequence data from standard metagenomic libraries using bioinformatic tools. However, high-throughput strategies are yet to be developed to access the total TE population present in a community.

11.3. VIRAL METAGENOME

Bacteriophages also transfer genetic material between bacterial hosts and, though they are not MGE in the same sense as plasmids,transposons or integrons, they may be considered as MGE. Metagenomic approaches to access the viral population from several microbial ecosystems have been studied nowadays. In future this will create a new revolution in virology.

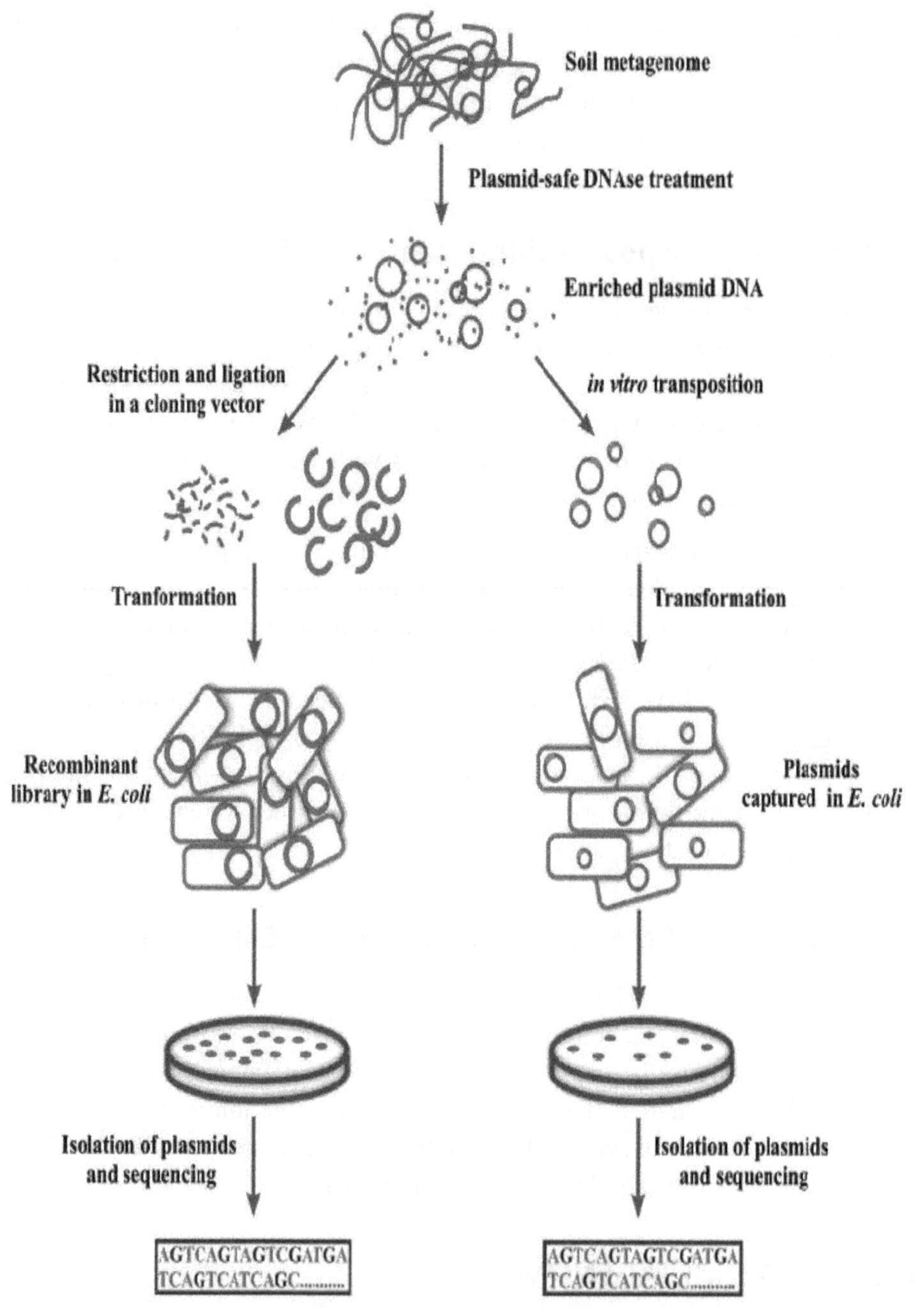

Fig. 11.1. Selective enrichment of plasmid metagenome. The plasmid-safe DNase selectively removes linear DNA fragments and enriches circular double-stranded plasmid DNA molecules. Shot-gun libraries are created with enriched plasmid DNA. Alternatively plasmids are captured by in vitro transposition. In transposon-aided capture (TRACA), native environmental plasmids are captured independent of their origin of replication and phenotypes.

In general, libraries of total viral DNA or cDNA extracted from a complete viral population are constructed. To access the phage metagenome, bacteriophage particles are first separated from the environmental samples and nucleic acids are then extracted. Isolating phage metagenome is complicated by the presence of free and cellular DNA in the environment. Therefore, a combination of differential filtration with tangential flow filters, DNase treatment and density gradient centrifugation in caesium chloride is used to separate the intact viral particles from the microorganisms and free DNA.

Once intact virions have been isolated, the viral DNA is extracted and cloned. Cloning of phage metagenomic DNA in standard *E. coli* strains is also challenging due to the use of modified bases in phage genomes, which is essential to resist host restriction endonucleases, and the presence of lethal genes such as holins and lysozymes. Amplification of the phage DNA by PCR based strategies or whole genome amplification (WGA) can be employed to produce copies of phage DNA lacking modified bases, and to generate sufficient amounts of DNA for making libraries. Viral metagenomics and phage metagenomics are discussed in detail in next chapters.

11.4. RELATIONSHIPS BETWEEN METAGENOMES AND MOBILE ELEMENTS

A metagenome does not necessarily reveal the phenotype of the source community because the expression and interactions of the genes are dependent on their arrangements with in genomes. The basic units of organization in genomes are operons in microbial genomes and modules in phages (Figure 11.2). Although technically different, these units (or gene clusters) are similar, relatively autonomous and can move around while retaining their functionality. Distinct microbial and phage genomes are assembled by mixing and matching these clusters *via* homologous recombination. Interaction rates between different operons and modules are increased by professional mobile elements, including plasmids, pathogenicity islands and temperate phages (represented by prophages within microbial genomes). Unlike operons and modules, MORONs and ORFans are single genes that become incorporated into phage or microbial genomes *via* illegitimate recombination. Usually these sequences are lost through deletion, but occasionally these genes are maintained by positive selection (e.g. specialization genes). The ecological and evolutionary barriers between phage and microbial metagenomes are blurred by these horizontal gene transfer events.

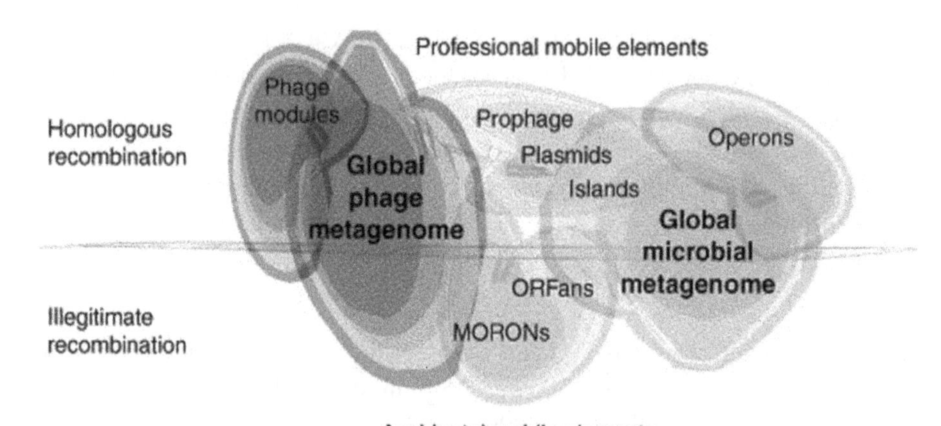

Fig. 11.2. Relationships between metagenomes, operons, modules and mobile elements.

11.5. ACCIDENTAL MOBILE ELEMENT

Metagenomes are populations of genes within a particular sample. The basic arrangement of these genes is modules in phages and operons in microbes. Professional mobile elements facilitate the movement of modules and operons between and within genomes. Any gene can potentially become an accidental mobile element *via* illegitimate recombination. Examples of accidental genes include specialization genes like exotoxins, *psbA* and *phoH*.

The mobile genetic elements are enormously applied in metagenomic studies. The viral metagenomics is dealt in a separate chapter.

Mini Quiz

1. What is TRACA and its significance?
2. Define transposable elements?
3. Explain the selective enrichment of plasmid metagenome?
4. Elucidate the correlation between metagenomes and mobile elements?
5. Explain the term accidental mobile element?

CHAPTER - 12

VIRAL METAGENOMICS - GENERATION OF VIRAL METAGENOME

Viruses are the most abundant and genetically diverse biological entities on Earth and the vast majority is yet to be discovered. Therefore, systematic surveillance for viruses requires techniques that have both broad specificity and high sensitivity. Conventional laboratory techniques in virology often fail to detect a specific etiology in many syndromes that are thought be caused by viruses. Metagenomics-based tools such as pan-viral microarrays and ultra-high-throughput sequencing have significantly improved our ability to detect and characterize divergent as well as novel viruses. Some of these methods rely on the fact that any one virus will possess some degree of conservation within its genomic sequence with other members of the same family. Thus, nucleic acid amplification tests targeting conserved regions in the viral families associated with a particular disease can often lead to a successful diagnosis. However, metagenomics-based techniques such as pan-viral microarrays are able to transcend our predetermined lists of viruses associated with each syndrome and allow for the simultaneous interrogation of thousands of conserved and specific genetic regions within all taxa of known virus families. Second generation high-throughput sequencing offers the unique opportunity to discover novel pathogens with no *prior* sequence information with sensitivities comparable to that of PCR. As the costs for these techniques continue to decrease and the technology becomes more widely available, they will have the potential to revolutionize our approach in detecting viruses and diagnosing viral diseases.

This chapter deals with the metagenomic approaches to study environmental viral diversity and the potential influence of viral movement on evolution is discussed in the context of microbial metagenomes.

12.1. ABUNDANCE AND IMPORTANCE OF VIRUSES IN THE ENVIRONMENT

The discovery that there were millions of microbes in every milliliter of seawater and gram of sediment, begged the question, 'Who is eating them?' Early studies showed that protists, primarily nanoflagellates, were important grazers of marine microbes. Viruses entered the picture in 1989, when Bergh *et al.* used transmission electron microscopy to show that there are ~10 million virus-like particles (VLPs) per ml of sea water. At the time, however, this observation did not make sense because the vast pool of environmental microbes had yet to be discovered. Most environmental VLPs are assumed to be phages, viruses that infect bacteria, because bacteria are the most common prey items. In addition, viral abundance often correlates strongly with microbial abundance, with a fairly constant ratio of 5–10 VLPs per bacteria. Further studies showed that phages kill between 4–50% of the bacteria produced every day, with the rest being eaten by protists. Apart from this VLPs play a major role in horizontal gene transfer and are acting as a catalyst for microbial evolution.

12.2. CULTURE-INDEPENDENT STUDIES OF VIRAL DIVERSITY

Several characteristics make viruses well suited for high throughput sequencing and analysis. For example, the hardiness of many viral capsids makes them ideal for concentration and purification. The small size of viral genomes, particularly of phages, can be advantageous for many bioinformatic techniques and full genome assembly.

12.2.1. Generation of metagenome of a viral consortia or virome

Metagenomics is the sequencing of all nucleic acids isolated from an environmental sample; therefore, any remaining cells, nuclei and free nucleic acids will contaminate the resulting virome. Eukaryotic and most microbial genomes are significantly larger than viral genomes (with a few exceptions, such as Mimivirus and Mamavirus) and can provide a disproportionately large amount of the sequence data within a virome. Therefore, the eukaryotic and microbial cells are destroyed before viral nucleic acid extraction. Free nucleic acids can also contaminate viral metagenomes. DNase I treatment before capsid lysis can reduce, but not completely eliminate free DNA. Some RNA viruses contain RNA in the nucleocapsid structure. RNase treatment to rid samples of free and cellular RNA may therefore result in the loss of those viral particles. After the nucleic acid extraction steps, routine testing for the presence of contaminating cellular DNA should be conducted by attempting to PCR-amplify small subunit rRNA (ssuRNA) genes using universal primers for 16S and 18S. As viral genomes do not contain either16S or 18S genes, the presence of these sequences will provide a semi-quantitative measure of microbial and eukaryotic contamination within the virome.

The steps involved in generating viral metagenome comprises of isolation, concentrating and viewing viral particles.

12.2.1.1. *Concerns for virome generation*

1. The composition of the sample will dictate the viral concentration and extraction protocol.
2. The amount of sample required depends on the ease of the viral recovery.. For a standard pyrosequencing metagenome (approximately 1–5 mg of DNA), requires isolation of ~3 $x10^{11}$ viruses.
3. Virions are lost at every step; viral particles can become immobilized on filters, destroyed in storage and adsorbed to larger particles.
4. Viruses can also be enveloped and/or have various modifications to their structures

Hence the overall strategy should aim to ensure adequate viral-particle concentration and elimination of contaminating cells from a variety of sample types

12.2.1.2. *Tangential-flow filtration*

Tangential-flow filtration has been used to isolate viral particles from a variety of environments. Particles smaller than the filter pore sizes are pushed out through the filters (filtrate) (Fig. 12.2). A back pressure is used to force the filtrate through the holes. The remaining sample (retentate) is then collected into a reservoir basin and repeatedly cycled through the filters. Recirculation therefore concentrates the large particles. Reservoirs can be a variety of basins, such as 1-liter glass Erlenmeyer flasks or 10-, 20- and 80-liter-size plastic carboys. These reservoirs should be sterilized before use.

1. Sample pretreatments

e.g., prefilter with Nitex, homogenize, reduce sulfide bonds

2. Concentrate VLPs

e.g., tangential-flow filtration, PEG precipitation, Microcon

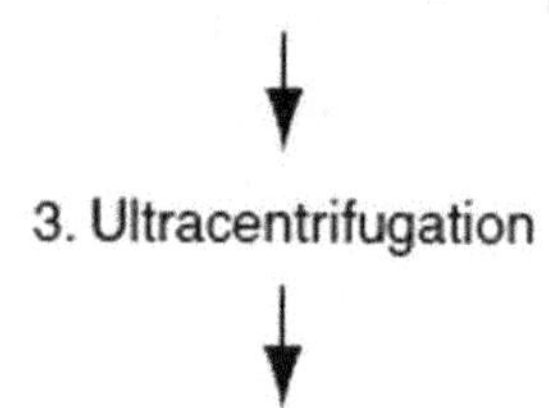

3. Ultracentrifugation

4. Extract nucleic acids

DNA: formamide/CTAB
RNA: modified RNeasy Kit

6. Amplify nucleic acids if necessary

DNA: Phi29 DNA polymerase
RNA: Transplex RNA amplification

7. Sequencing

Fig. 12.1. Flowchart for viral metagenome generation.

Beneficial aspects of this filtration approach are

(i) the large surface area that allows large volumes of filtrate to pass through rapidly and (ii) the tangential flow that prevents clogging of the system (e.g., as with an impact filter).

(a) The original sample is passed through the filter using a peristaltic pump and pressure gauge. The retentate is recirculated to concentrate microbes (A retentate).

(b) The filtrate containing the viruses is expelled (A filtrate) or saved for further viral-particle concentration using a 100-kDa filter. Viral particles are concentrated in the final 100-kDa retentate.

Tangential-flow filtration filters are available in several pore and surface area sizes depending on the purposes. *Viz.*, 0.2-mm (Fig. 12.2a) and 100-kDa filters (Fig. 12.2b). Tubing is attached to either end of the TFF (see Fig. 12.2b), and the sample is run from the collection reservoir through a peristaltic pump and a pressure gauge and then across the TFF, concentrating the microbes (0.2 mm; Fig. 12.2a) or microbes and viruses (100-kDa TFFs; Fig. 2b) in the 'retentate.' It is important to always keep the pressure within the tubes less than 10 p.s.i. (~62 kPa), as higher pressures can compromise the viral particles. If samples contain particulate matter that can clog the filter, first pass the sample through a Nitex mesh (e.g., pore size ~100 mm). Avoid exposing filters to organic solvents (e.g., chloroform), as these compounds will ruin the filters.

12.2.1.3. Polyethylene glycol (PEG) precipitation

Polyethylene glycol precipitation can be incorporated if sample volumes are too large and need to be concentrated even further before CsCl centrifugation or DNA extraction. This is often done when large viral lysates or filtrates are greater than 50–100 ml after being filtered or when diluted for DNase treatment.

12.2.1.4. Purification of viral particles by CsCl density centrifugation

Viral particles can be purified using density gradient ultracentrifugation, which is based on the physical properties of various virions. The solvent and speed of the centrifugation, as well the types and number of gradient layers chosen, is entirely dependent on the density of target viruses. It is also important to note that gradients should be made from the same buffer as the samples (e.g., PBS), and that this buffer has been purified with a 0.02-mm filter to ensure that external viruses do not contaminate the resulting fractions. For example, for seawater phages, density layers of 1.7, 1.5 and 1.35 g $ml^{-}1$ CsCl are prepared using 100-kDa filtered seawater (Fig. 12.3). The sample is also brought up to 1.12 g $ml^{-}1$ CsCl and then loaded onto the heavier CsCl layers and centrifuged at 22,000 r.p.m. (~60,000g to 82,000g maximum) at 4 °C for 2 h in a swinging bucket rotor.

In contrast, due to the lower buoyant density of many human associated viruses, 1.7, 1.5 and 1.2 g ml^{-1} layers are made in 1x PBS.

(a) The CsCl density gradient is poured by adding 1 ml of the heaviest density (left arrow) and then carefully floating the next density (right arrow) by tilting the tube and slowly adding in the lighter density.

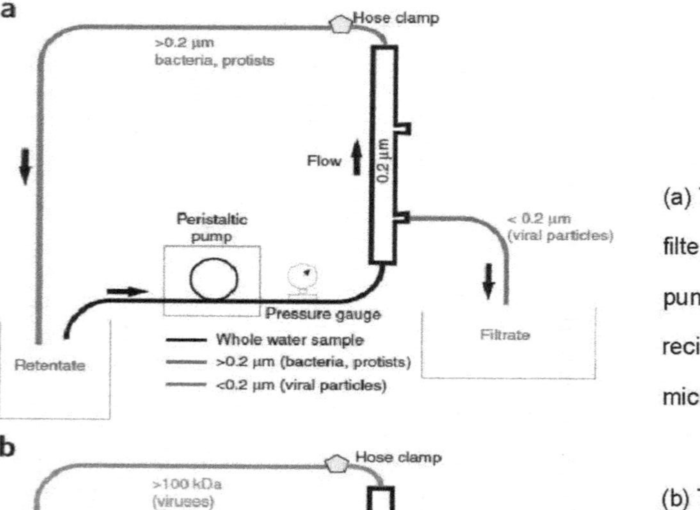

(a) The original sample is passed through the filter using a peristaltic pump and pressure gauge. The retentate is recirculated to concentrate microbes (A retentate).

(b) The filtrate containing the viruses is expelled (A filtrate) or saved for further viral-particle concentration using a 100-kDa filter. Viral particles are concentrated in the final 100-kDa retentate.

Fig. 12.2. Tangential-flow filter setup for viral-particle concentration.

(b) Phages are concentrated in a four-layer gradient. The resulting gradient localizes the phage between the 1.5 and 1.35 g ml^{-1} densities. A sterile 18-gauge needle is inserted just before the 1.5 g ml^{-1} density to remove the concentrated phages (see arrow).

(c) After centrifugation, virions are extracted by placing the tube over a reservoir and piercing the tube with an 18-gauge needle and sterile syringe.

(d) The fraction just above the 1.7 g ml^{-1} step is removed (marker line).

For making gradients, different researchers prefer different pouring methods (Fig. 12.3a). Pipetmen and tips can be used for pouring every density layer. However, graduated pipettes can also be used for either some or all of the layers. In addition, syringes and needles can be used to pour the layers. Generally, CsCl layers are added first with pipetmen and then the sample with a graduated pipette by slowly dripping the sample along the side of the column so that it does not disturb the lightest density. It is extremely important not to disturb the boundaries

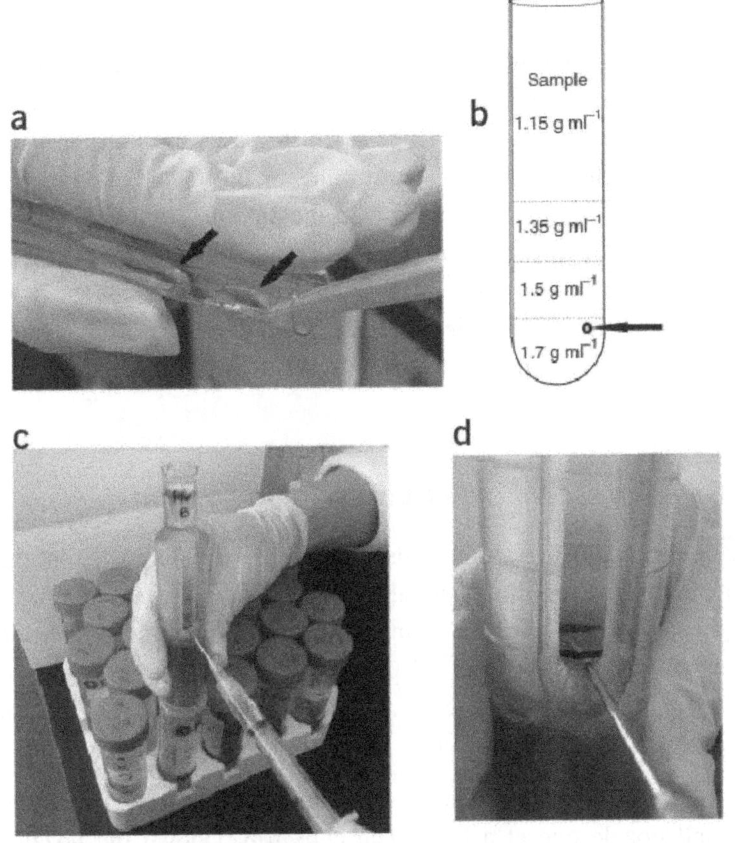

Fig. 12.3. Buoyant density centrifugation setup and methods. Viruses are isolated using different densities of CsCl and ultracentrifugation.

of the layers before centrifugation. To practice accurate density layer formation, use different CsCl densities labeled with a different food coloring. This makes it easy to visualize layer mixing. Also, although the amount of sample added will depend on the size of the tube and rotor used, it is important to completely fill the column tube, or it can collapse during centrifugation.

This protocol is specifically designed for phage particle concentration. These viral particles are located at 1.5–1.35 g ml^{-1} in CsCl density boundaries after ultracentrifugation

12.2.1.5. *Cell sorting*

In addition to being able to analyse the properties of cells, some flow cytometers are able to sort specific cell populations of interest. These pure populations of cells can then be used in future experiments, cultured or re-stained with other fluorescent dyes and re-analysed. In most cell sorters, cells are passed in a stream of fluid out through a narrow orifice, at which point they pass through a laser beam and are analysed in the same way as in a standard flow cytometer. A vibration is passed to the sample stream, which causes it to break into droplets at a stable break off point. If a cell of interest passes through the laser beam it is identified and when it reaches the droplet of the break off point an electric charge (positive or negative) is applied to the stream. As the droplet leaves the stream it passes through deflection plates carrying a high voltage and the droplet will be attracted to one of these plates, depending on the charge it was given. Uncharged droplets pass through undeflected and deflected droplets are collected in tubes. In this way two different populations of cells can be sorted from the one sample.

12.2.1.6. *Fluorescence Activated Cell Sorting (FACS)*

In multicellular organisms, all the cells are identical in their DNA but the proteins vary tremendously. Therefore, it would be very useful if we could separate cells that are phenotypically different from each other. In addition, it would be great to know how many cells expressed proteins of interest, and how much of this protein they expressed. Fluorescence Activated Cell Sorting (**FACS**) is a method that can accomplish all these goals.

The process begins by placing the cells into a flask and forcing the cells to enter a small nozzle one at a time. The cells travel down the nozzle which is vibrated at an optimal frequency to produce drops at fixed distance from the nozzle. As the cells flow down the stream of liquid, they are scanned by a laser. Some of the laser light is by the cells and this is used to count the cells. This scattered light can also be used to measure the size of the cells.

If you wanted to separate a subpopulation of cells, you could do so by tagging those of interest with an antibody linked to a fluorescent dye. The antibody is bound to a protein that is uniquely expressed in the cells you want to separate. The laser light excites the dye which emits a color of light that is detected by the photomultiplier tube, or light detector. By collecting the information from the light (scatter and fluorescence) a computer can determine which cells are to be separated and collected.

12.3. VERIFICATION OF VIROME PURITY

Column filtration (Fig 12.4) and epifluorescence microscopy (Fig12.5) are used to verify that viral samples do not contain contaminating nuclei or microbial cells.

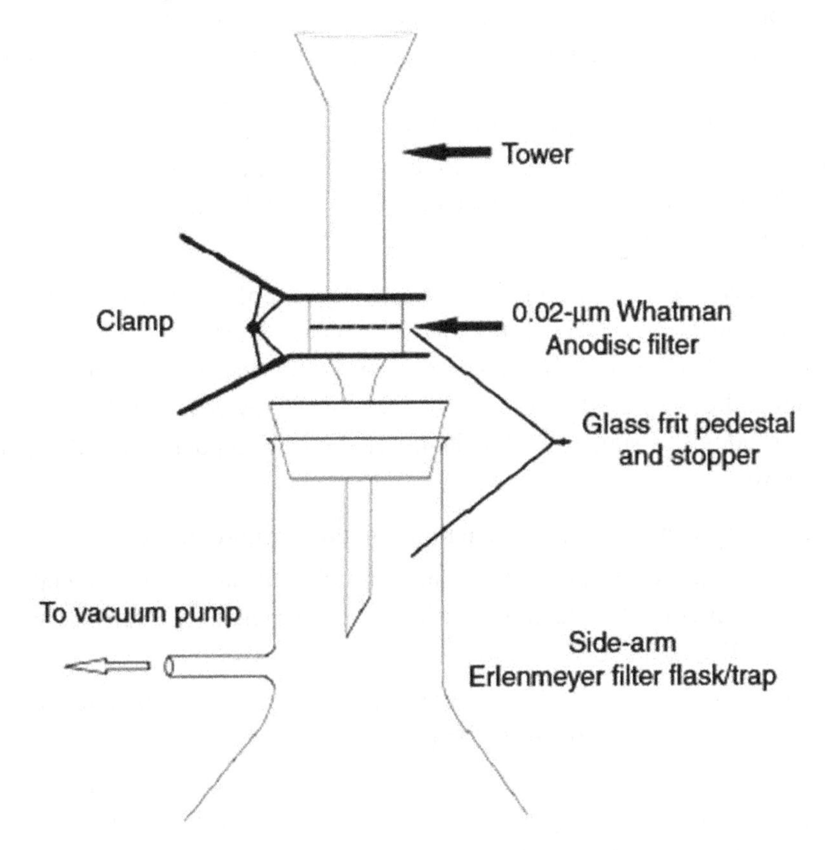

Fig. 12.4 . Filtration setup for viral enumeration and verification of sample purity.

In brief, aliquots of the viral concentration layers are diluted in 5 ml of buffer solution and then slowly (<10 p.s.i. or ~62 kPa) filtered through 0.02-mm filters. Filters are then removed from the tower glass frits and stained with SYBR Gold. The precise dilution of the SYBR dye is at the user's discretion. Concentrations greater than 1X will result in brighter and longer-lasting slides but is more expensive and may result in higher background staining. This nucleic acid stain will work on both live and fixed samples. If working with live samples, it is critical to use an isotonic buffer instead of water in the protocol.

For fixing samples, prepare a fresh solution of 4%(wt/vol) paraformaldehyde in buffer (e.g., 4% paraformaldehyde in sterile seawater) and then dilute the sample with an equal volume of fixative, to give a final concentration of 2%. Viral-particle enumeration has been shown to be significantly affected by the storage time and fixation method used. It is therefore absolutely critical that fixed sample slides are made as quickly as possible (preferably immediately) after viral concentration. SYBR Gold is viewed in the FITC channel, excited with blue light and fluorescent in the green. Viruses appear as pinpoints of light on unconcentrated water samples (Fig.12.5a, arrows). Nuclei and microbial cells are much brighter and larger (Fig. 12.5c,d, arrows). For samples with large volume, it is easy to overload a filter with viruses. If the background appears milky or grainy, then it is most likely that the volume of viral concentrate added to the slide was too high (Fig. 12.5d). Dilute another aliquot and remake the slide until individual virions are visible (Fig. 12.5b).

To verify that only virus particles were collected from TFF and/or ultracentrifugation, construct filtration device. Glass frit and filter pedestal is placed within an Erlenmeyer filter flask/trap. Trap is attached to vacuum pump. Viral 0.02-mm filter is placed atop glass frit. The glass tower is attached to the glass frit pedestal with a stainless steel clamp. The sample is poured into the graduated glass tower.

(a) Seawater sample with characteristic pinpoint fluorescence.

(b) Viral particles appear as dim pinpoints of light next to larger and brighter microbial cells. Concentrated virions and microbial cells (arrow) from 100-kDa TFF retentate.

(c) Virus concentrate after CsCl ultracentrifugation containing contaminating eukaryotic and microbial debris (arrow).

(d) Fluorescence image showing the characteristic milky appearance of a filter overloaded with virus particles. All images were taken at 600x magnification with an oil immersion objective.

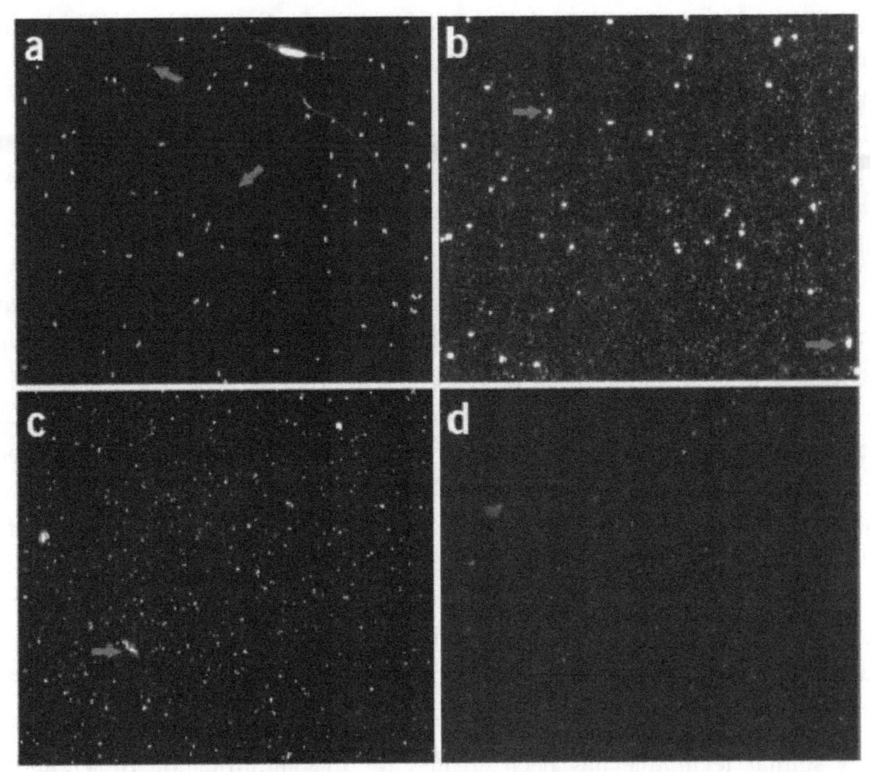

Fig. 12.5. Epifluorescent microscopy to determine viral-particle concentration and purity.

12.4. NUCLEIC ACID EXTRACTION

Viral DNA extraction combines the formamide/cetyltrimethylammonium bromide (CTAB) or phenol/chloroform method. Although a large majority of our work is on DNA viruses, metagenomes have been generated for RNA viruses, particularly from medical samples and seawater. For sequencing, sufficient quantities of cDNA from the viral RNA should be generated. When isolating RNA viruses, always use RNase-free and viral-free solutions. Also, some viruses incorporate RNA into their coat structures, and treatment with RNase will destroy those particular RNAs. In such cases, it may be preferable to accept some level of contaminating host RNA.

12.5. NUCLEIC ACID AMPLIFICATION

If adequate quantities of DNA cannot be prepared directly from the samples, the DNA can be amplified using multiple displacement amplification. This method makes use of Phi29 DNA polymerase, an enzyme with strand-displacement activity that enables it to amplify genomic DNA using random primers and a single denaturation step. A single 20-ml Genomiphi reaction (containing 10–100 ng of template DNA) will produce ~4 mg of amplified DNA. DNA produced using this technique has been shown to be underrepresented in the 500–1,000 bp near the termini of linear templates. To reduce sequence bias and ensure high yield, it is recommended that several independent amplification reactions be conducted for each sample.

Recent reports have also suggested that two reactions be performed for each sample, one with and one without the denaturation step to reduce certain types of sequence bias. The amplification can be verified by running an aliquot on a 1% (wt/vol) agarose gel or by taking an optical density reading on a spectrophotometer. The optical density ratio of A260/A230 indicates nucleic acid purity, and should ideally be in the range of 1.8–2.2. Values less than 1.0, often obtained with environmental samples, indicate the presence of contaminants that may interfere with enzymatic reactions. However, for an optical density measurement to reflect polymerized DNA, it is necessary to repurify the DNA by using silica columns intended for genomic DNA so that primer oligomers and dNTPs are removed. The amplification process not only increases total DNA abundance, but also effectively purifies the DNA.

A similar technique has been created for the amplification of RNA. One particular method is called TransPlex for Whole Transcriptome Amplification of Intact and Degraded RNA. This kit can amplify total RNA in less than 4 h without any terminal bias. The RNA template is primed with quasi random primers. Reverse transcriptase and displaced strands are used as additional templates for further amplification.

Mini Quiz

1. What is VLP? How does it act as a catalyst for microbial evolution?
2. Define virome and explain the concerns of virome generation?
3. Detail the methods to isolate viral particles from a diverse environment?
4. What is the principle of CsCl density centrifugation?

CHAPTER - 13

MAPPING THE VIRUS WORLD THROUGH METAGENOMICS

Viral abundance in the environment exceeds that of bacteria. Viruses have a significant influence on microbial communities. Molecular analysis is essential to investigate the diversity of viral assemblages because the majority of viruses are uncultured due to a lack of suitable hosts, such as bacteria. Indeed, the cultivation of viruses that infect eukaryotes is not easy. A PCR-based approach is also not appropriate because there are no universally conserved genes or markers for viruses like the16S rRNA gene for bacteria. Whole viral assemblage genome sequencing (viral metagenomics) recently overcame these limitations and became a promising method by which to investigate uncultured viral diversity. By this approach, viruses are purified and concentrated by sequential filtrations and ultracentrifugation and whole viral genomes are extracted, amplified, and sequenced by shotgun cloning or pyrosequencing. The advantages of sequence-independent amplification and metagenome sequencing for characterizing novel viruses are that they are simple, fast, and without bias toward any particular viral group.

13.1. CULTURE-INDEPENDENT STUDIES TO EXPLORE NOVEL VIRUSES-WHY?

13.1.1. Conserved gene studies of viral diversity

Studies of specific microbial species in the environment were severely limited by culturing bias until Pace *et al.* introduced 16S rDNA analyses. The fact that most microbes have not been cultured has severely limited studies of viral diversity. As a further complication, there is not a single genetic element shared by all viruses, meaning that it is not possible to study total viral diversity using 16S rDNA-like approaches. However, whole genome comparisons have shown that there are conserved genes shared amongst all members within certain viral taxonomic groups. These conserved genes (or signature genes have been used to study diversity within known groups of viruses among cultured isolates, as well as in the environment.

The diversity of phages infecting cyanobacteria, for example, has been studied extensively using sequences of structural proteins. The diversity of algal viruses and T_7-like Podophages (a group of genetically related, tailed, dsDNA phages) has been examined by sequencing DNA polymerase genes, and a RNA-dependent RNA polymerase fragment has been used to identify at least two novel groups of marine picorna-like viruses that probably infect eukaryotic algae. The vast majority of sequences found in these studies belong to novel subgroups that are not represented by cultured isolates. All of the conserved gene studies suggest that *environmental viral diversity is high and essentially uncharacterized.*

13.1.2. Metagenomic studies of Viral diversity

Studies employing conserved genes do not enable the discovery of completely novel groups of viruses. Aquatic and soil environments consistently contain 10^7–10^8 viral particles per milliliter of water or gram of soil, respectively, and there are thousands of viral types in each sample. To assess total dsDNA viral diversity, partially sequenced shotgun libraries from uncultured viral communities is essential. Approximately 75% of the sequences in these viral metagenomes did not match any genes in the database suggesting that most viral diversity remains uncharacterized.

For instance, in an investigation with viruses from fecal matter and marine environment, using the sequences that could be associated with a particular phage group, demonstrated that certain phage groups (T_7-like Podophage, ›-like Siphophage, and T_4-like Myophage) were abundant in all the marine communities, whereas other, unrelated groups were more common in fecal matter. The phage groups most abundant in fecal matter are known to infect Gram positive bacteria, whereas the marine phage groups infect Gram-negative bacterial hosts. This suggests that the most abundant phage groups in a given environment reflect the kinds of microbes found in that environment.

Sequencing viral metagenomic libraries also provides information about the underlying community structure (i.e. the number of genotypes and relative abundances. Occurrence of the same DNA sequence in different shotgun fragments means that the same viral genotype has been sampled multiple times. To take advantage of this information, a modified version of the Lander-Waterman algorithm was developed to model uncultured viral communities based on metagenomic data. These analyses show that uncultured viral communities are some of the most diverse communities ever observed. There might be, for example, up to a million different viral genotypes in one kilogram of marine sediment. Together, these culture-independent studies of viral diversity show that the majority of viral diversity remains unknown, with most viruses belonging to novel groups without any cultured representatives.

13.1.3. Movement of viruses between biomes

If all environments have unique, endemic viral populations, then extrapolation from metagenomic data predicts global viral diversity to be ~100 million distinct viral genotypes. Alternatively, viruses might be moving between environments. In this case, local diversity could be quite high but global diversity would be relatively limited. To differentiate between these two hypotheses, conserved genes have been used to show that identical, or nearly-identical, phage-encoded sequences are present in different biomes. These phage-encoded sequences are so similar that they must have moved between environments within recent evolutionary time.

For example, one phage encoded DNA polymerase sequence, named HECTOR, was found in marine water, soil, rumen fluid, associated with corals and in solar saltern water. Assuming an average burst size of 25 particles and an average half-life for phages of 48 h, a phage encoding the HECTOR sequence would need to complete a lytic cycle once every ten days (i.e. five half-lives) to survive. This phage would complete ~36 generations per year. The mutation rate for dsDNA phage-sized genomes is 10^{-7}–10^{-8} changes bp^{-1} per generation. Therefore, it is expected that there would be a 1-bp change in the HECTOR sequence every ~525 years. The HECTOR sequence never differed by >3 bp over a 533-bp fragment, therefore, this sequence has moved between these environments within the last 1000–2000 years. It is not known whether a complete HECTOR-encoding phage, or just this piece of DNA, is moving between environments (e.g. as suggested by the mosaic model.

It is evident from many studies that viruses can find hosts in different biomes. Two possible explanations for this observation are: (i) identical microbial hosts are found in the different environments; or (ii) viruses are not completely host-specific and can attack the different microbial hosts found in each environment. Currently, the second possibility seems more likely because cross infecting viruses are relatively common. If most viruses viruses actually do attack multiple hosts, this will dramatically reduce estimates of global viral diversity and change models of virus–host dynamics.

13.1.4. Viral community structure

Viruses are both globally distributed and have high diversity on the local scale. In the bank model, only the most abundant viruses are active. The rest of the viruses are inactive, rare and form a potential population for recruitment, much like a seed bank in plant populations. This distribution matches the observed rank-abundance curves predicted by modeling of marine metagenomic libraries, where the most abundant viral genotype comprises <5% of the total community and the vast majority of viral genotypes are extremely rare (<0.01% of the community). When an environment changes, different hosts grow and the viruses preying on these hosts move from the 'bank' into the 'active' fraction. The previously abundant and active viruses start to decay and enter the bank fraction. Then Bank model predicts that the viruses in the active fraction should stay the same as long as the host populations remain susceptible to these viruses. Changing the active host population is predicted to cause a rapid change in the active viral fraction. Supporting this model, culturing and genome size distribution studies provide evidence that specific viruses become abundant, decay to undetectable levels and then return on seasonal cycles.

In the Bank model, the active viral–host pairs are behaving in a 'Kill-the-Winner' fashion where the most abundant host population is reduced by its viral predators. Population reduction of the most abundant host creates an open niche, which enables a new host to become abundant. In this manner, the identity of the most abundant host is constantly changing. The Bank and Kill-the-Winner models only vary in their treatment of the inactive fraction. However, data from some viral decay studies show that virions display a rapid initial decay, which is followed by a period of slower decay, possibly because the remaining viral particles are more refractory to harmful environmental factors. The viruses in the bank fraction were probably previously active and are now in various states of population decline. Alternatively, the rare viruses in the bank fraction might be supplied via constant production of a low level of virions (e.g. induction of proviruses) or allochthonous inputs.

13.1.5. PHAGE SPECIALIZATION TO DIFFERENT ENVIRONMENTS

In general, officially recognized phage taxa share similar suites of genes. These genes are usually grouped together into modules and are likely optimized to work together. Examples of modules are the virion assembly genes or DNA replication machinery. It is relatively easy for phages to mix and match these modules and produce viable recombinants.

Phage modules are 'fuzzy' entities and they often have other genes inserted into them. These genes have been termed MORONs (for more DNAs) and they can come from other phages or hosts. Insertion of these extra DNA elements can occur *via* different illegitimate recombination mechanisms. Over evolutionary time, MORONs are deleted or they can be maintained by positive selection. In the second case, the behavior of the phage and its host can be dramatically affected. A well known example of this phenomenon is the acquisition of exotoxin genes by phages. These exotoxins convert the phage-infected microbes into pathogens.

Phages adapt to new environments by acquiring ecologically important genes as MORONs. The cyanobacteria *Synecococcus* and *Prochlorococcus* are major marine autotrophs. Photosynthesis by these two genera account for approximately one-third of the carbon fixed in the marine environment. Cyanophages infecting *Synecococcus* and *Prochlorococcus* have acquired genes involved in photosynthesis. The *psb*A gene, which encodes the D1 protein, has been found in all three of the completely sequenced *Prochlorococcus* phage genomes and one *Synecococcus* phage genome . D1 is a rate limiting photosynthesis protein and host-derived D1 concentrations dramatically drop during phage infection. Expression of the phage-encoded *psbA* gene enables the phage to maintain photosynthesis throughout the infection cycle, presumably providing the phage with energy. In the open ocean, the energy benefit associated with the acquisition of photosynthesis genes by cyanophages was probably a key evolutionary step.

Similarly, phosphate is a major limiting nutrient in parts of the ocean and many marine phages encode enzymes involved in phosphate metabolism (e.g. phoH, RNA reductases, and endonucleases). The photosynthesis and phosphate metabolism genes are embedded into phage genomes that have clear relationships to well-studied phages that infect *Escherichia coli.* The take-home message is that phages acquire ecologically important genes to adapt to new environments. Phages can carry these genes between environments, contributing to local and global lateral gene transfer.

13.2. VECTOR –ENABLED METAGENOMICS (VEM) FOR DIVERSITY OF PLANT VIRUS

Current knowledge of plant virus diversity is biased towards agents of visible and economically important diseases. Less is known about viruses that have not caused major diseases in crops, or viruses from native vegetation, which are a reservoir of biodiversity that can contribute to viral emergence. Discovery of these plant viruses is hindered by the traditional approach of sampling individual symptomatic plants. Since many damaging plant viruses are transmitted by insect vectors, "vector-enabled metagenomics" (VEM) was developed to investigate the diversity of plant viruses. VEM involves sampling of insect vectors from plants, followed by purification of viral particles and metagenomic sequencing. The VEM approach exploits the natural ability of highly mobile adult insects/other vectors to integrate viruses from many plants over time and space, and leverages the capability of metagenomics for discovering novel viruses. VEM successfully characterizes the active and abundant viruses that produce disease symptoms in crops, as well as the less abundant viruses infecting adjacent native vegetation.

13.3. FUTURE PROSPECTS OF VIRAL METAGENOMICS

Although local viral diversity is extremely high, viruses appear to be moving between environments, which constrains total global viral diversity and provides a conduit for horizontal gene transfer. Mathematical modeling from metagenomic data suggests that most viral genotypes are relatively rare, serving as a 'Bank' of viruses that can become abundant when their hosts become dominant (e.g. in response to changing environmental conditions).Armed with new techniques for detecting and enumerating viruses, these ubiquitous creatures have been found in a wide range of environments, including the deep sea, solar salterns (which are ten times saltier than the ocean), acidic hot springs (>80°C with pH=3.0) , alkaline lakes (pH=10, under>30 m of ice in polar lakes and in the terrestrial subsurface (>2000 m deep).

Future research needs to examine the relationships between uncultured viruses found in different environments, as well as the spatial and temporal scales on which viral communities change. More than simply providing sequence data, metagenomic analyses can offer insights into biogeographical distributions, community structure and ecological dynamics. In addition, the crucial genes for microbial adaptation to a given environment are likely to be moved by phages and can therefore be identified by analyzing the genomic content of viral communities. Continued metagenomic studies of viral communities will enable the estimation of global viral diversity, as well as a deeper understanding of the impact of horizontal gene transfer on microbial diversity and evolution.

Mini Quiz

1. What is HECTOR?
2. Define MORONS?
3. Explain virus enabled metagenomics and its role in plant virus diversity?
4. Name the signature gene used for viral diversity studies?

CHAPTER - 14

FUNCTIONAL VIRAL METAGENOMICS

The enzymes of bacteriophages and other viruses have been essential research tools since the first days of molecular biology. However, the current repertoire of viral enzymes only hints at their overall potential. The most commonly used enzymes are derived from a surprisingly small number of cultivated viruses. The extreme abundance and diversity of viruses revealed over the past decade by metagenomic analysis enables to access the treasure trove of enzymes hidden in the global virosphere and develop them for research, therapeutic and diagnostic purposes.

14.1. VIRAL ENZYMES AND THE DEVELOPMENT OF MODERN BIOTECHNOLOGY

Molecular biology research always depends on enzymes for almost every manipulation used in amplification, detection, cloning, expression, mutagenesis and analysis of nucleic acids. Although many cellular enzymes have been used in these methods, viruses, including bacteriophages, have been an especially rich source of useful enzymes. Phages particularly T4, T7, lambda, M13 and phiX174 were the first model systems of molecular biology. T4 phage is still the most prolific source of useful viral enzymes. Its 169 kb genome and estimated 300 genes include at least 85 genes involved in DNA replication, recombination and repair, and nucleotide metabolism, transcription and translation.

14.1.1. Viral genomes as instrumental agents- why?

Enzymes from viruses and other phages have been instrumental in the development of the field of biotechnology. One reason is the density of certain genes in viral genomes. For example, a typical bacterial genome of about 2 Mb contains only a single *pol* I gene (coding for DNA polymerase I). By contrast, between 20 and 40 pol genes per 2Mb were found in viral metagenomic sequences.

Viral genomes are relatively simple compared with those of their hosts, and contain a comparatively high proportion of genes coding for structural proteins (e.g. for coat and tail) together with proteins involved in nucleic acid metabolism and lysis. In contrast to the extreme diversity and abundance of viruses in the environment the most common viral enzymes used today are derived primarily from a small number of phages (e.g. T4, T7, lambda, SP6 and phi29) and retroviruses (e.g. Moloney murine leukemia virus (Mo-MLV) and avian myeloblastosis virus (AMV)].

14.1.2. Viral enzymes as research reagents

Many research applications of viral enzymes are centered on nucleic acid metabolism (Table 14.1). DNA polymerases (Pols) are the main focus of much of the discovery efforts. These enzymes are essential for common molecular biology techniques including whole genome amplification, PCR,Sanger (dideoxy chain termination) DNA sequencing, and most methods for nucleic-acid-based detection of infectious agents, cancer and genetic variation. Most of the nextgeneration sequencing platforms (e.g. Roche/454, Illumina, Helicos, Pacific BioSystems) use multiple microbial and/or viral DNA Pols both for template preparation and base discrimination.

14.1.2.1. *Bacterial Pols vs Viral Pols*

Viral DNA Pols are functionally distinct from their cellular counterparts. All of the microbial DNA Pols used as reagents are derived from two families: bacterial Pol I and archaeal Pol II, which are highly similar cellular repair enzymes and not true replicases. By contrast, viral Pols are highly diverse in terms of primary amino acid sequence and biochemical activities. As replicase enzymes, they have distinct properties.

a) *phi29 Pol: phi29 Pol* has a processivity of >70 000 nucleotides (i.e. it incorporates over 70 000 nucleotides before dissociating), far greater than that of *Thermus aquaticus* (Taq) Pol, with only 50-80 nucleotides. Additionally, phi29Pol has a strong strand-displacement capability, which, along with its high processivity, makes it the polymerase of choice for whole genome amplification by multiple displacement amplification (MDA).

b) *T7 phage Pol:* T7 phage *Pol* holoenzyme has a processivity of >10 000 nucleotides and efficiently incorporates chain terminating nucleotide analogs; these attributes made it an optimal choice for Sanger sequencing until it was displaced by *Thermosequenase*, a Taq Pol derivative that was engineered on the basis of sequence features in T7 DNA Pol that conferred efficient incorporation of dideoxynucleotides.

c) *T5 phage Pol:* T5 *Pol* has both high processivity and a potent strand-displacement activity, which are independent of additional host or viral proteins.

d) *T4 phage Pol:* The DNA Pols of T4-family phages have high proofreading activities that are commonly exploited for generating blunt ends, especially in physically sheared DNA.

e) *Pyrophage Pol:* Many uses for DNA Pols including PCR, RT PCR, thermocycled Sanger sequencing and certain whole genome amplification methods depend on thermostability up to 95°C. Before metagenomic screens were used to discover the so-called PyroPhage Pols, no known viral Pol could withstand this temperature.The power of viral metagenomics is illustrated by the discovery of novel classes of thermostable viral DNA polymerases. Viral particles were isolated from hundreds of liters of water from two Yellowstone hot springs (74°C and 93°C) and separated from microbial cells by tangential flow filtration. Viral nucleic acid (<100 ng) was extracted from each sample, sheared, amplified and cloned. In collaboration with the US Department of Energy Joint Genome Institute, about 29 000 sequence reads were determined (approximately 28 Mb in total). The sequences contain several hundreds of apparent pol genes, including one from every known pol family. Only 59 of these genes were full length, and ten genes were expressed to produce thermostable DNA Pols (PyroPhage Pols). The predicted amino acid sequences of seven PyroPhage Pols were compared with representatives of known viral and cellular Pol families and with the commonly used thermostable Pols. The PyroPhage Pols fall into two groups distinct from other known viral and microbial Pols, one group much more diverse than the other.Consistent with the molecular diversity, the biochemical activities of the PyroPhage enzymes is also unique. For instance, PyroPhage 3173 Pol possesses inherent reverse transcriptase activity and the highest thermostability of known viral Pols, which allow its use in conventional PCR and single-enzyme RT PCR. It also shows promise in whole genome and single cell amplification.

14.1.2.2. *Other viral enzymes*

i) *RNA polymerases (RNAPs):* RNA polymerases (RNAPs) transcribe RNA from a DNA template. In contrast to their cellular counterparts, the RNAPs of phages T_7, T_3 and SP6 function as independent proteins that recognize short promoter elements (of 17 nucleotides in length) without the requirement of transcription factors or other cellular components. They can generate large amounts of RNA for direct use, or for in vitro or in vivo protein synthesis. RNAPs are also key components of several transcription-mediated amplification approaches.

ii) *DNA ligases:* Virtually all ligation methods used for cloning and linker attachment depend on T_4 DNA ligase, because of its relatively high activity on 50 and 30 extended and blunt DNA.

iii) *Integrases and recombinases:* The integrases and recombinases of various phages have been used to integrate genes into the genomes of a wide variety of bacterial and eukaryotic cells. The lambda-red-mediated system is used for stable transgene integration into the *Escherichia coli* genome whereas the phage P1 cre/lox system functions in mammalian, fungal, plant and other eukaryotic cells.

iv) *Resolvases:* Resolvases (e.g. T4 endonuclease VII and T7 endonuclease I) have been used to detect single nucleotide polymorphisms (SNPs) .

v) *Replicases:* The replicases (reverse transcriptases) of retroviruses,especially M-MLV and AMV, are used for reverse transcription of RNA to form cDNA; they are indispensable for research on transcription processes and RNA viruses and for transcriptome analysis.

vi) *DNA methylases:* DNA methylases are gaining significance in studies of epigenetics; most reported methods use bacterial methylases, but genes encoding other methylases are common in viral genomes

vii) *Transposases:* Transposases are used for insertional mutagenesis and nested deletions, and as an alternative to subcloning or primer walking for sequencing longer templates. Although bacterial transposases are more commonly used, phage Mu transposase has also been used for insertional mutagenesis and sequencing.

viii) *Enzymes from thermophilic phages:* Genes encoding useful enzymes have been discovered recently in viral genomes outside of the usual core group of phages and retroviruses.

   a) The RNA ligase from a *Thermus scotoductus* phage, for example, is ten times more efficient at joining single-strand DNA molecules than is the T4 RNA ligase.

b) The thermophilic phage GBSV1 encodes a nonspecific nuclease useful for degrading RNA and single- and double-stranded, circular or linear DNA.

Table 1. Current and future uses of viral enzymes.

Enzyme	Viral source(s)	Current and emerging application(s)
DNA polymerase	T4, T7, phi29, PyroPhage Pol	Conventional and next-generation sequencing,amplification, end repair
Reverse transcriptase M-MLV,AMV, RNA polymerase	PyroPhage RT, T7, T3, SP6	cDNA cloning, microarrays, transcriptome analysis Probe generation, in vitro expression, molecular diagnostics, isothermal amplification
RNA replicase	Q beta, phi6	RNA amplification, production of siRNA
DNA ligase	T4	Cloning, linker ligation
RNA ligase	T4, TS2126	Joining DNA and/or RNA
DNA repair	T4	Mutation detection, dermatology
Polynucleotide kinase	T4, RM378	End repair, end labeling
Transposase	Mu	*In vivo* mutagenesis, genomics
Helicase	None known	Improved amplification and sequencing
Recombinase	Lambda Red	*In vivo* recombination
Integrase	Lambda (Int/att), P1 cre/lox	Site-specific recombination
Methylase	None known	Genomics, epigenetics
Nonspecific nuclease	T7, lambda	Cloning, nucleic acid removal
Resolvase	T4, T7	Mutation detection
RNase H	T4	cDNA synthesis, mutation/SNP detection, isothermalamplification
Lysozyme	T4	Protein/plasmid isolation, antimicrobial, diagnostic
Tail protein	Many sources	Bacterial typing, bacteriostatic
Saccharolytic enzyme	Many sources	Biofilm remediation
Protease	TEV	Site-specific cleavage
Coat protein	SSV	Nanocompartments, imaging, drug delivery, internalcontrols for reverse transcriptase PCR

c) Phi6 replicase is used for replicating RNA, particularly to generate small interfering RNA (siRNA) and micro RNA (miRNA) for RNA interference studies.

14.2. CLINICAL APPLICATIONS OF VIRAL ENZYMES

In addition to research uses, several viral proteins are useful either as therapeutics or as diagnostic reagents.

14.2.1. Tail proteins in diagnostics

Sensitivity to phage infection is often the only means of distinguishing between closely related bacterial strains. For instance, pathogenic *Bacillus* strains are distinguished using a strain-typing phage. But it is a time-consuming process. To overcome this bottleneck, more direct and rapid tests are developed based on direct detection of binding of differentially labeled phage tail proteins.

14.2.2. Lysozymes and tail proteins as antimicrobials

Phage proteins are potential therapeutic agents in fighting against infectious bacterial diseases. Though the interest in phage therapy declined with the discovery and development of antibiotics, the emergence of multi-drug resistant bacterial pathogens is reviving interest in phage therapy for both humans and agricultural species. The whole phage particles are used as antimicrobial agent in phage therapy. But the isolated phage lytic enzymes are highly specific antimicrobial agents that often target gram-positive pathogens without affecting beneficial co-occurring organisms. Some viral tail proteins also have bacteriostatic activity and might be useful as antimicrobial compounds.

14.2.3. Repair enzymes as anticancer agents

T4 endonuclease is a DNA repair enzyme with activity against ultraviolet-induced cyclobutane pyrimidine dimers. When applied to the skin, this enzyme has shown protective properties against damage from exposure to the sun, significantly reducing the incidence of basal cell carcinomas and actinic keratoses. T4 endonuclease is useful for patients with DNA repair deficiencies such as xeroderma pigmentosa. Other viral enzymes that reverse alkylation of RNA bases also find clinical applications.

14.2.4. Saccharolytic enzymes as treatments for recalcitrant infections

The most intractable bacterial pathogens are present in the environment within biofilms adhered to a solid surface. Microbial biofilms in biomedical devices such as catheters and implants are particularly problematic as they can interfere with the device or cause sepsis, and are highly resistant to antibiotics. Some phages express saccharolytic enzymes that degrade biofilm carbohydrates and are useful in treating pathogens present in biofilms on medical implants or responsible for recurrent infections.

14.2.5. Coat proteins in imaging, drug delivery and production of diagnostic standards

The viral coat proteins have the ability to self-assemble around other molecules. This property has been exploited to encapsulate imaging agents and drugs. For example, encapsulation of contrasting agents assists in magnetic resonance imaging. This is useful for targeted delivery of anticancer and antimicrobial drugs. The ability to engineer binding domains allows for specific targeting of cancer or microbial cells.

14.3. METAGENOMIC APPROACHES TO DECIPHER NOVEL VIRAL PROTEINS

Technical challenges related to the cultivation of new viral–host systems are the primary impediment in discovering new viral enzymes. Traditional approaches require that both the host and the virus must be amenable to cultivation. Hosts that fail to form lawns and viruses that fail to form plaques can preclude isolation of the virus. Once isolated, significant empirical work is required to define parameters such as multiplicity of infection (MOI), burst size, and infection kinetics. These factors are important for detecting viral proteins that are induced at specific time points after infection. It is often difficult to discern the viral proteins from those of the host cell. Isolation of genes from metagenomic sequence libraries circumvents many of the limitations in cultivation-based discovery and is a valuable alternative approach to the discovery of novel enzymes. However, attempts at mining functional enzymes using metagenomic approach are only just in its budding stage.

14.4. CHALLENGES IN METAGENOMICS-BASED ENZYME DISCOVERY

The focus of viral metagenomics shifts from purely ecological investigations of diversity to uncover novel biological features of viruses and ultimately, producing useful enzymes. The practical challenges arising and the means to overcome some are discussed below:

a) **Conventional clone libraries:** Next-generation sequencing instruments, including the Roche/454 FLX pyrosequencer, the Illumina Genome Analyzer and ABI Solid have exponentially increased the amount of information that can be extracted from a sample of DNA or RNA. However, with the recent introduction of the 454 Titanium pyrosequencer, sequence read lengths are just beginning to be adequate to allow occasional retrieval of entire genes from single reads. Although the avoidance of a clone library step in these next-generation sequencing platforms has significant advantages in terms of cost and speed. The lack of clone library makes retrieval of specific sequences for follow-up functional characterization more difficult.

   Traditional random "shotgun" clone libraries combined with Sanger sequencing have advantages for functional metagenomic investigations. First, the longer reads often contain entire coding sequences. Second, the clones that are generated during the process can be fully sequenced or directly expressed to produce enzymes. In fact, with sufficiently large inserts (3–5 kb), it has been possible to isolate multigene operons containing functionally related genes.

b) **Library construction:** An inherent difficulty in viral metagenomics is isolating adequate amounts of genomic material for large-insert library construction. Although viruses are abundant, their genomes are generally < 50 kb in length. A liter of water with 10^9–10^{11} viral particles or a kilogram of soil often yields sub-nanogram quantities of genomic DNA or RNA. Because a typical microbial genome can be equivalent in size to hundreds of viral genomes combined, a few microbial cells or low amounts of free nonviral DNA can represent significant contamination in a viral nucleic acid preparation. This Contamination issues are addressed through a combination of filtration, centrifugation and nuclease treatments which are employed in retrieving viral particles from hot springs (Figure 14.1).

   Non-sequence-specific amplification allows production of metagenomic libraries from the resulting viral DNA. The quality of both the amplification and the library construction are crucial to the success of a metagenomic expression screen. The difficulty of library preparation increases with longer average

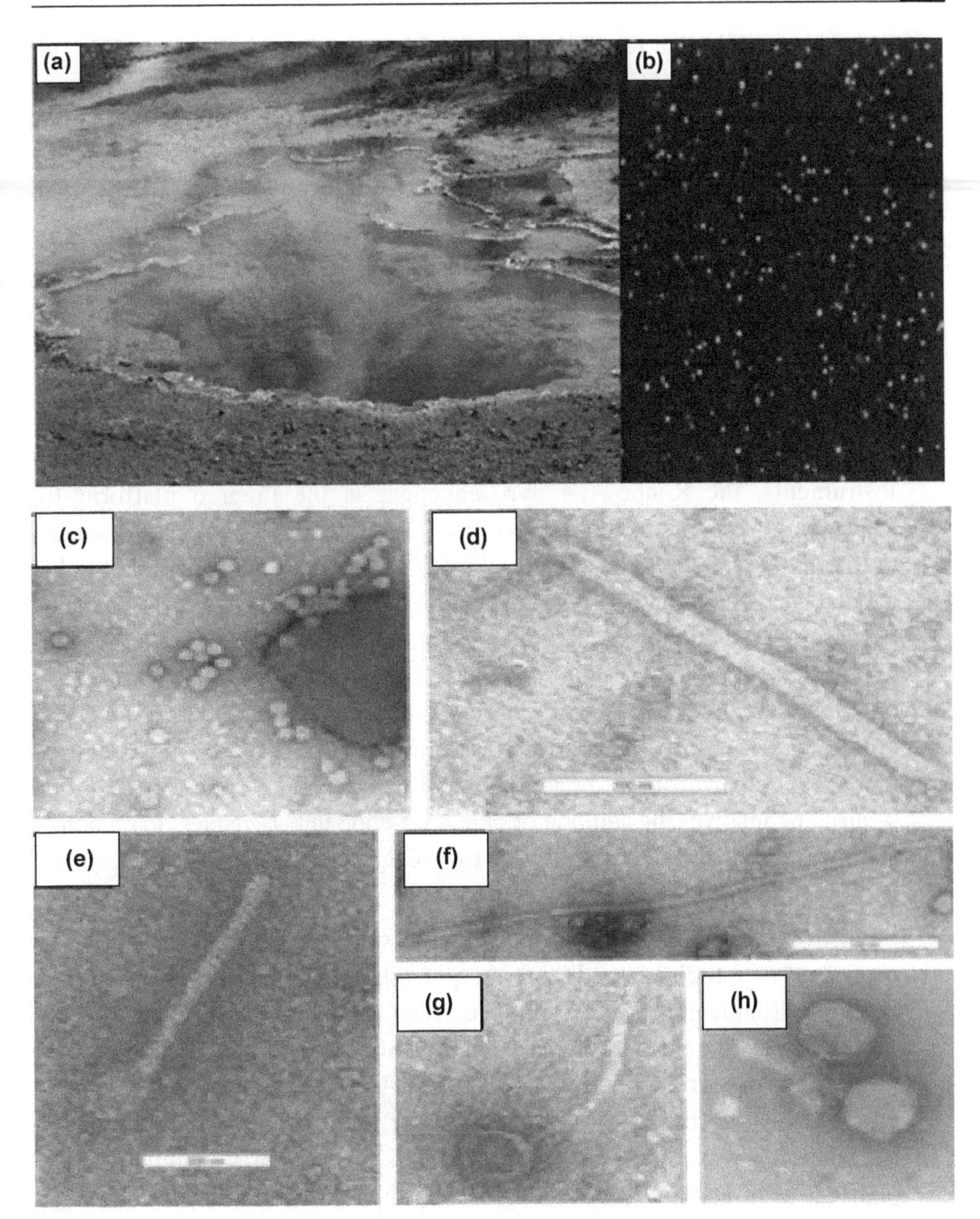

Figure 14 1. Viruses enriched from natural environments, such as (a) hot springs, by filtration and differential centrifugation. Enrichments are imaged by (b) epifluorescence microscopy to quantify the viral particles and to ensure the absence of contaminating microbial cells.Transmission electron micrographs of viral preparations – as these shown from (c, d, e) the Firehole River and (f, g, h)White Creek areas of Yellowstone National Park –allow determination of viral types based on morphology *[Courtesy: Trends in Microbiology*].

insert sizes and, as a result, most viral metagenomic shotgun libraries have insert sizes of <3 kb. Cloning artifacts or chimeras complicate expression of authentic genes, and sequence stacking caused by amplification or cloning biases reduces the comprehensiveness of the screen. Viral genomes are difficult to clone without significant bias using standard high-copy-number vectors with leaky promoters (e.g. pUC19), compared with transcription-free circular or linear cloning vectors. Most work has been focused on DNA viruses because the genomes of RNA viruses must be reverse transcribed before cloning. Nevertheless, RNA viral libraries have been constructed and sequenced which are a valuable source of enzymes such as reverse transcriptases, RNA replicases and proteases.

c) **Next-generation sequencing:** Of the next-generation DNA sequencing instruments, the Roche 454 pyrosequencer is the favored platform for sequencing viral metagenomes because of its relatively long read lengths (currently around 400 nucleotides). Viral DNA is amplified using phi29 Pol in a method called multiple displacement amplification (MDA) before sequencing because the technique requires microgram quantities of DNA. Although this approach is rapid, MDA is known to have significant amplification bias when applied to small starting amounts of bacterial genomic DNA and will preferentially amplify circular DNA. Because of these biases, the preferred method for amplification of environmental viral DNA before next-generation sequencing involves the addition of oligonucleotide linker-adapters to randomly fragmented viral DNA, followed by PCR amplification using primers homologous to the adapter sequences

d) **Sequence assembly:** In viral metagenomic studies, sequence assembly has been used as a tool for predicting the potential genotypic diversity within a given environmental viral assemblage. Very large scale sequencing of viral metagenomes should permit the assembly of large contiguous stretches of DNA and potentially entire genomes. The high degree of sequence polymorphism within viral populations assembles large contigs with high confidence. The estimates of genotypic diversity relies on assemblies of 95% match over at least a 20 bp overlap. Such high stringency tends to prevent misassembling noncontiguous parts of the genome. However, these assemblies probably overestimate the amount of ecologically meaningful population diversity within a given viral assemblage. Alterations in the assumed size of the average phage genome within a viral assemblage or changes in the percentage match used in the assembly can have dramatic effects on resultant diversity estimates.

Moreover, the genomes of closely related viral types can diverge significantly and assembly at high stringencies fails to associate reads from related, but genetically distinct viral types. Such stringent assembly criteria can lead to overestimation of unique viral types, and can prevent discovery of genes, enzymes and genomes. For example, lowering the assembly stringency of two high temperature viral communities from 95% to 50% resulted in the assembly of operons and potentially entire genomes from an environmental viral sample. Although these lower stringency assemblies are probably composites of partial sequences with inherent microheterogeneity, they provide insight to the possibly dominant viral populations within a given hot-spring environment. Unfortunately, synthesis of a gene from such a low stringency assembly is probably not a viable strategy for accessing new enzymes because the inherent polymorphism prevents accurate translation and ultimately proper folding of an expressed protein. This can be circumvented by accessing the full insert of the original constituent clones or by developing PCR primers to selectively amplify, clone and sequence the putative genes from the original viral DNA preparation.

e) **Identifying genes:** Owing to the high diversity of viral genes and the relatively low numbers of viral genomes in public sequence databases, most viral coding sequences (even for well-studied phages such as the coliphages T4 or T7) have no significant similarity to any known genes. For most viral metagenomic data with relatively long reads, 30% or fewer of the sequences show similarity to a previously identified gene, and many of these genes are not associated with an identifiable function. The rate of sequence similarity is substantially lower for short read sequences. The likelihood of finding genes by similarity also depends on the evolutionary conservation of the genes.

For example, genes coding for lysozymes are highly conserved and commonly detected, whereas those coding for holins are highly diverse and are virtually undetectable in viral metagenomes. In some cases, a sizable proportion of viral metagenomic sequences (up to 60%) show homology to environmental sequences, indicating some level of prevalence and persistence of viral genes across a variety of environmental contexts. Thus, improved understanding of the phylogeny and ecology of viral genes will come from a more thorough bioinformatic exploration of predicted viral proteins encoded by metagenomic libraries. A thorough knowledge of viral biology improves annotation of viral genes and the utility of metagenomics as a predictive tool for the potential influence of viral processes within microbial communities, and enhance enzyme discovery.

f) **Expression of the discovered genes for enzyme production:** Similarity-based detection is only useful for enzyme discovery if the respective genes can be expressed in a suitable host. This depends on accurate determination of the start and stop sites, a process that is hindered by characteristics unique to metagenomes in general and to viral metagenomes in particular. A basic problem is that most ORF prediction programs were written for complete cellular genomes. Even after assembly, metagenomic sequences tend to be short compared with the whole genomes for which programs such as Glimmer and Gene-Mark were developed. The flood of metagenome sequence data has spurred the development of new ORF calling algorithms, such as MetaGene Annotator, designed specifically for ORF prediction from fragmentary sequence data. But the applicability of these newer ORF prediction algorithms for downstream expression studies is not available so far.

Other challenges in expressing viral genes derive from unique characteristics of viral biology. The commonly faced problems in producing functional proteins from viral metagenomes are:

i) Nonstandard codon usage
ii) Involvement of cellular proteins and post-translational processing
iii) Overlapping genes especially in RNA and single-stranded DNA viruses
iv) Codon usage bias

For example, in vivo expression of M-MLV reverse transcriptase involves recoding of the cellular translation system to read though a stop codon that produces a polyprotein (Gag–Pol fusion) must be post-translationally processed to form an active enzyme. The erroneous assignment of the start codon of the T5 phage DNA Pol gene, turned out to be a rare TTG start codon which hampers expression.This mistake was discovered by determining the amino terminal sequence of the phage T5 native protein; however, such an approach is impossible for proteins discovered through metagenomic analysis because phage isolates are not available for purification of native proteins.

14.5. MITIGATION STRATEGIES

i) **Post translational processing:** In post-translationally processed polyproteins, the amino and/or carboxy terminus of the functional protein is determined by protease recognition sequences within the protein sequence or by sitespecific autolytic cleavage and not by the start or stop codons of the genes. Some polyproteins appear to be capable of autoproteolytic cleavage in vitro to form functional proteins, whereas in other cases, it is possible to insert artificial start codons based on alignment to known genes and produce functional proteins.

ii) **Codon bias:** The problem of codon bias can be overcome by expressing genes in cells that supply tRNAs for rare codons (e.g. Rosetta cells) or by resynthesizing the gene with optimal codons for expression in *E. coli.*

14.6. FUNCTIONAL SCREENING OF VIRAL METAGENOME

Functional screening is an alternative approach to sequence-based discovery that is not dependent on initial DNA sequencing. Rather, clones of environmental DNA are screened directly for enzymatic activity. This approach selects for genes that can be detectably expressed in the host, usually *E. coli*. Functional screening requires an assay capable of screening a large number of clones with few false positives or false negatives. Plate-based colony assays are typically the simplest screens, but they are not amenable to detection of certain useful proteins. Alternatively, colonies can be picked and screened by highthroughput robotic methods. The random nature of shotgun libraries means the coding sequence might not necessarily be in proximity to a vector-based promoter, especially if a transcription-free vector is used. Therefore, a gene of interest must express from its own promoter, which might not be active in the host cell at a detectable level. An advantage of functional metagenomics-based enzyme discovery is that metagenomic screens are well suited to focusing on specific environments, which might result in discovery of enzymes with desirable attributes.

In addition, target genes might have unstable secondary structures when cloned or their encoded proteins might be toxic to the host cells, preventing their isolation. Despite these challenges, two significant advantages make functional screens worth considering.

a) First, this approach can identify genes that are too divergent from known genes to be identified by sequence similarity.
b) Second, once clones are detected by function, the genes can be expressed immediately (at least at low levels) to produce proteins that can be further characterized.

14.7. FUTURE OF VIRAL METAGENOMICS

As we add new and unique enzyme activities to the "tool chest" of molecular biology, several sources should be explored. The viral metagenomic studies have Shown that the pool of sequenced viral genomes is woefully unrepresentative of extant viral diversity. Thus, viral metagenomics offers a means of exploring genetic diversity within the vast uncultivated portion of the virosphere. However, practical application of functional viral metagenomics as an approach to new enzyme development is in its infancy. Functional metagenomics promises to provide a wealth of starting information for the development of enzymes applicable to a broad range of industrial, biomedical and research applications.

Viral metagenomics promises to feed an almost unlimited diversity of enzymes into screens that can be tailored to the practical needs of a variety of applications. For instance, hot springs have proved a fertile source of thermostable enzymes whereas sewage effluents would seem to be likely sources of phage-derived enzymes specific for lysis of human enteric pathogens. Screening viromes from a broad range of environments will certainly provide a vast reservoir of genetic diversity for discovery of the next generation of molecular tools and medicines. More efficient approaches based on advances in genomics and screening technology will accelerate this work. The short read lengths of ultrahigh-throughput next-generation sequencing technologies are currently problematic for viral metagenomics- based enzyme discovery. However, these problems should be surmountable, and the advantages that these methods offer in throughput should make them a logical choice for functional metagenomics. Read lengths from the current generation of 454 pyrosequencing instruments are approaching those of Sanger shotgun sequencing. In the near future, single-molecule methods might allow sequencing of entire or nearly entire viral genomes in a single read obviating the need for assembly of sequences or recovery of clones. Alternatively, some of the singlecell genomics methods might be applicable to viral genomes. Using either approach, the ability to determine entire sequences of genes will allow direct to synthesis of genes. Moreover, such long read sequencing will allow for the isolation of entire coding sequences including complete multigene operons.

Another promising technology is micro and nano-fluidics which should improve our ability to perform ultrahigh-throughput functional screens. With the availability of these technologies, remaining challenges will be: (i) improving gene predictions to relate evolutionarily distant proteins to a possible function, (ii) developing the informatics to sift through the huge amounts of data to find these genes and (iii) improving our ability to rapidly express and characterize the new enzymes.

Mini Quiz

1. Discuss the advent of viral enzymes and their impact in modern biotechnology?
2. Differentiate between bacterial Pols and viral Pols?
3. Narrate the metagenomics approaches to decipher novel viral proteins?
4. What are the approaches for functional screening of viral metagenome?

CHAPTER - 15

METAGENOMICS AND INTEGRATIVE 'OMICS' IN BIOREMEDIATION

Our planet suffers more and more from various pollution problems. Global anthropogenic pollution has led to the accumulation of a wide variety of organic and inorganic xenobiotic moieties causing detrimental effects on human health and pristine ecosystems. Microbial communities are known to colonize at contaminated sites and have the ability to metabolize these recalcitrant xenobiotic anthropogens. In the current scenario, several microbial bioremediation strategies are sought after as an indispensable, ecofriendly and cost-effective solution towards restoring the polluted ecosystems. Microbial bioremediation strategies can be either applied ex situ or in situ in order to restore a contaminated environment.

Ex situ treatment involves removal of contaminants at an off-site separate treatment facility usually a bioreactor or an effluent treatment plant (ETP), whereas *in situ* treatment involves on-site bioremediation of the contaminated site either by natural attenuation (progressive removal of contaminant), intentional biostimulation of indigenous microbial communities by providing electron acceptors/ donors/nutrients or by bioaugmentation which includes deliberate inoculation of laboratory grown microbial strains having exceptional ability to degrade pollutants at the contaminated sites. Success of ex situ or in situ bioremediation relies heavily on the relative abundance, structure, catabolic versatility and biotic/abiotic interactions of the microbial communities (aerobic/ anaerobic) that are indigenously present, amended or stimulated at contaminated sites, industrial waste-water/ effluent treatment plants or within biofilms of the bioreactors.

Advent of molecular technologies over the past few decades has led environmental microbiologists to recognize microbial communities as an imperative ecological parameter in monitoring polluted sites either by detecting community shifts in response to pollution or their resilience towards anthropogenic disturbances. Cultivation-independent analyses of the overall microbial community structures at contaminated sites using molecular profiling techniques have been instrumental in our understanding of the community dynamics, relative abundance and distribution of micro- organisms actively involved in bioremediation. The intrinsic microbial communities underlying bioremediation at contaminated sites is depicted in Fig. 15.1. The application of these technologies for studying microbial communities and their functional roles in environmental bioremediation is discussed in this chapter.

15.1. MOLECULAR SURVEYS OF MICROBIAL COMMUNITIES AND THEIR FUNCTIONAL GENES AT POLLUTED SITES

It is well known in microbial ecology that cultivation-dependent approaches are unable to assess or access the uncultivable microbial majority. Hence, cultivation-independent microbial community profiling using rRNA genes is a preliminary step before metagenomic analysis. Apart from the phylogenetic "gold standard" 16S rRNA and 18S rRNA genes, other catabolite specific genes have also been used to detect functional capabilities of indigenous microbial communities in pollutant biodegradation. Making a census of microbial communities at contaminated sites has enables to gain in-sights of specific microbial groups that are sensitive or most affected, resilient and pre-dominant, or actively involved in bioremediation The methods used in direct extraction of mRNA and DNA from soils and sediments are quite similar, however successful recovery of mRNA from soil particles requires extra caution and care, owing to the poor stability of mRNA.

One of the key components within biological wastewater treatment plants (WTPs) is the microbial biomass within the activated sludge. Activated sludge differs from soil and sediments in many aspects due to its high biomass density, low humic acid content, and the presence of bacterial aggregate flocs. High-quality "community mRNA" (metatranscriptome) from various environmental samples including wastewater sludge samples can be extracted using size separation on gel electrophoresis. This method involves constructing cDNA libraries and sequence analysis of cDNA clones. The inhibitor-free nucleic acids thus obtained can be utilized for subsequent analyses such as amplification cloning and sequencing rRNA genes, genotypic fingerprinting techniques, environmental genomics and transcriptomics analyses.

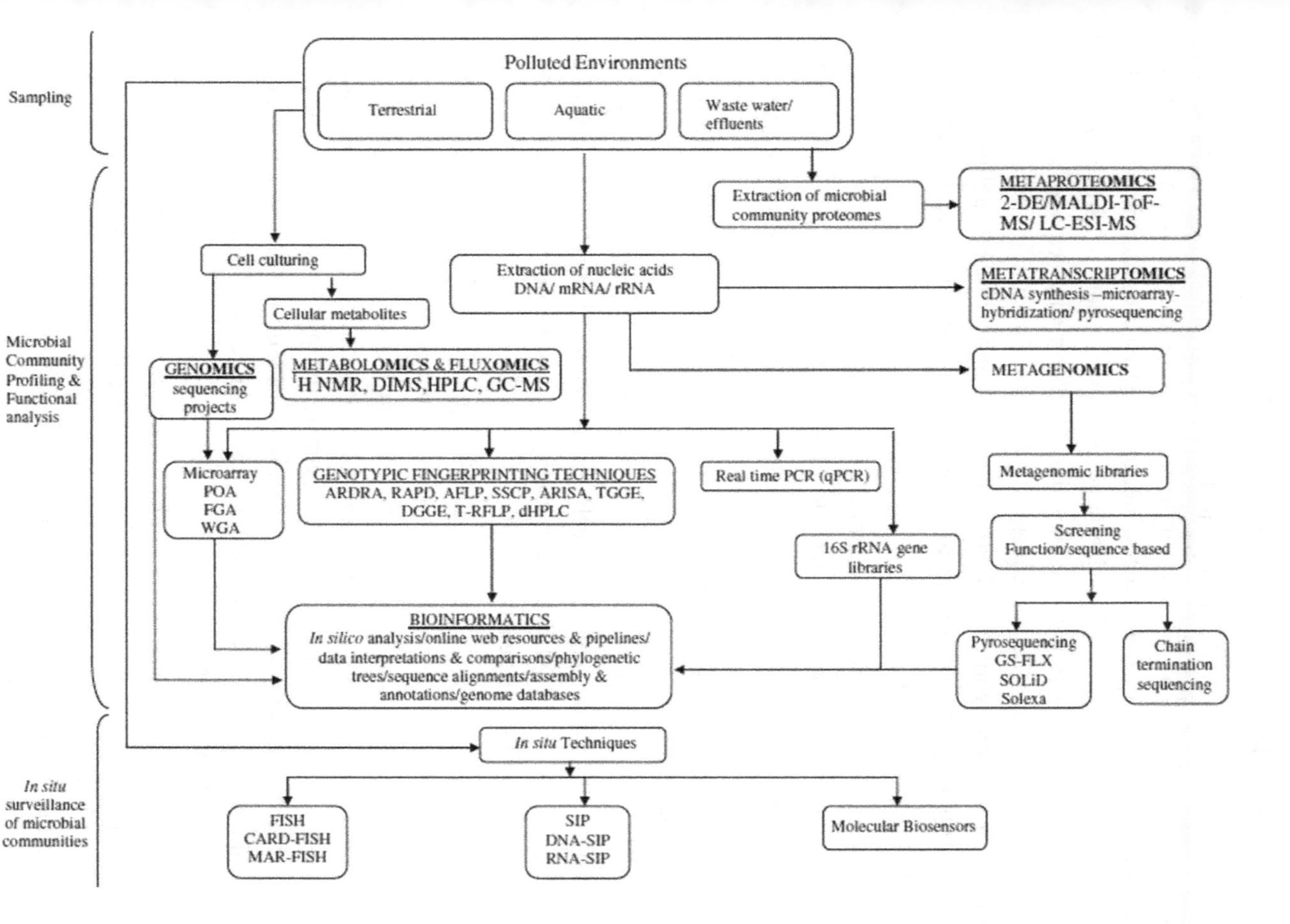

Fig. 15.1. An overview of technologies to survey microbial communities for bioremediation

15.1.1. SSU rRNA gene clone libraries, sequencing and multiplexed pyrosequencing

The classical approach of direct amplification, cloning and sequencing SSU rRNA genes from environmental DNA samples has completely redefined microbial ecological research over the past few decades and has resulted in accumulation of enormous amount of reliable rRNA gene sequence data. Characterization of microbial community structures, composition and shifts by sequencing of rRNA gene clone libraries has been frequently employed for studying bioremediation processes or identifying microbial bio-indicators of anthropogenic pollution. Some of the community changes observed in polluted site by sequencing 16S rRNA gene clone libraries include

i) Long-term influence of chromium pollution on soil microbial communities made a shifts from Proteobacteria to Firmicutes at chromium polluted sites.

   a) rRNA clones libraries: rRNA clones libraries have also been used to monitor active microbial populations involved in several bioremediation processes. Even though this standard approach of sequencing rRNA gene libraries provides most accurate and reliable census of microbial phylotypes, it is at the same time very cumbersome, time consuming and cost-intensive.

      Next generation sequencing technologies such as Roche 454 Life Sciences GS FLX or GS XLR based pyrosequencing are simplifying and rapidly changing this current scenario.

   b) Pyrosequencing: Pyrosequencing is a sequencing-by-synthesis method that is based on the detection of pyrophosphate (PPi) liberated during DNA synthesis using a cascade of enzymatic reactions and visible light detection modules such as photodiode, photomultiplier tubes, or a charge-coupled device (CCD) camera. This next generation pyrosequencing technology provides substantial cost reduction, ultra- fast sequencing, multiplex analyses of microbial communities and by-passes the requirement of cloning rRNA genes from environmental samples

   c) Genotypic microbial community fingerprinting techniques: A number of PCR-based genotypic fingerprinting techniques are available for monitoring microbial communities and efficacy of bioremediation processes, which can be listed as: amplified ribosomal DNA restriction analysis (ARDRA), randomly amplified polymorphic DNA analysis (RAPD), amplified fragment length polymorphisms (AFLP), length heterogeneity PCR (LH-PCR), single strand conformation polymorphism (SSCP), automated ribosomal intergenic spacer analysis (ARISA), denaturing gradient gel electrophoresis (DGGE)/

temperature gradient gel electrophoresis (TGGE) and terminal-restriction fragment length polymorphism (T-RFLP).

Recently, profiling of microbial community dynamics during ongoing bioremediation processes has enabled researchers to identify "key microbial strains" or "key catabolic genes" that are crucial for success of *in situ* or *ex situ* bioremediation. Consequently, these molecular techniques have become an important tool in management of microbial communities for defining optimal operational conditions and thereby increasing the efficacy of bioremediation processes.

i) Denaturing gradient gel electrophoresis (DGGE) and temperature gradient gel electrophoresis (TGGE): These are based on the principle of amplifying rRNA or functional gene PCR products from community DNA using primers containing a 50 bp GC-clamp and their separation on polyacrylamide gels having chemical or temperature based denaturing gradients. Microbial community shifts were detected in poly-metal contaminated soils (Li *et al.*, 2006) and abundances of dsrB (dissimilatory sulfite reductase b-subunit)-genes were assessed at an in situ metal precipitation site using DGGE fingerprinting technique (Geets *et al.*, 2006).

DGGE or TGGE have inherent limitations, since they are labor-intensive and often less reproducible in terms of band pattern and intensity detection obtained after electrophoretic separation. These limitations can be addressed by designing improved group-specific primers or by using a variant technique known as denaturing high performance liquid chromatography (dHPLC).

ii) Denaturing high performance liquid chromatography (dHPLC): (dHPLC) utilizes chromatographic separation instead of electrophoresis. In principle, the dHPLC technique separates a mixture of GC-clamped rRNA gene amplicons using an ion-pair reversed-phase HPLC under partly denaturing conditions. The resultant rRNA gene amplicon fragments are eluted from DNA-binding columns using triethy lammonium acetate buffers, the separated fragments (up to 1.4 Kb) can be assigned to peaks in the chromatogram and can also be re-amplified cloned and sequenced.

iii) Automated ribosomal intergenic spacer analysis (ARISA): It is based on the variation of species-specific amplicon lengths of the intergenic transcribed spacer (ITS) located between 16S and 23S rRNA genes, the technique involves amplification of ITS from the microbial community DNA using fluorescent labeled primers following the separation and detection of fluorescent ITS amplicons.

iv) *Length heterogeneity PCR:* LH-PCR is a yet another technique which involves detection of natural variations in lengths of different SSU rRNA genes within microbial communities.

v) *Single strand conformation polymorphism (SSCP)*: It is a fingerprinting technique that relies on the three-dimensional structure of single stranded rRNA genes which can be directly correlated with the polymorphisms present in the rRNA gene sequence. In this technique the electrophoretic mobility of single stranded rRNA genes is analyzed under nondenaturing conditions, and the resultant band patterns obtained are used to distinguish between microbial phylogroups.

vi) *Capillary electrophoresis single strand conformation polymorphism (CE-SSCP):* A recent advance in SSCP technique is known as capillary electrophoresis single strand conformation polymorphism (CE-SSCP). CE-SSCP involves amplification of rRNA genes from community DNA using fluorescent dye (e.g. 50-fluorescein phosphoramidite, FAM) labeled primers followed by separation of these mixed rRNA gene amplicons in the capillaries of automated sequencers (e.g. Applied Biosystems – ABI Prism 3100 Genetic Analyzer). The resultant amplicons are separated based on rRNA gene conformations produce CE-SSCP profiles of microbial communities that are detected by reading fluorescence signals as peaks.

vii) *Terminal-restriction fragment length polymorphism:* Terminal-restriction fragment length polymorphism analysis (T-RFLP) is yet another most popular technique in studying microbial communities. Main advantage of this profiling technique is its simplicity, automation and provision for accurate analysis of in silico data. T-RFLP analysis involves amplification of small subunit (SSU) rRNA genes from the total microbial community DNA using one or two primers tagged with different fluorescent labels. The resultant mixture of community rRNA amplicons is then digested with one or more four-base cutter restriction enzymes to generate T-RFs that are separated through capillary electrophoresis. Every single T-RF represents community fingerprints of a particular length unique to a particular phylogenetic microbial lineage or operational taxonomic unit (OTU). Based on the polymorphisms present in the SSU rRNA genes, different size of T-RF patterns of the whole microbial community are obtained. The size and relative abundances of these florescent T-RFs can be detected using an automated DNA sequencing instrument. Patterns of T-RF peaks on the output electropherogram can be identified using online database comparison tools.

Many online automated fragment length assignment tools such as (taxonomic assignment pipeline) TAP-TRFLP (http://rdp.cme.msu.edu), torast (http://

www.torast.de), and MiCA (http://mica.ibest.uidaho.edu), are available to perform in silico T-RFLP analysis of 16S rRNA gene sequences available in the databases. Similarly, an online tool known as TRF-CUT (www.arb-home.de) has been developed to predict in silico T-RFs using aligned small-subunit rRNA gene or functional gene sequences (e.g. pmoA, nirK, and nifH).

Unlike the above mentioned softwares, another online program TRiFle is available that can simulate and create T-RF datasets using arbitrary sets of DNA sequences from specific targets (e.g. genes involved in any metabolic pathways) or from unpublished sequences. These online predictive tools are useful in choosing appropriate combinations of primers and restriction endonucleases to achieve best resolution up to taxonomic level in T-RFLP analysis. Several web based tools such as phylogenetic assignment tool (PAT), TRUFFLER, APLAUS are available to determine microbial community composition by comparision with T-RFs predicted from an in silico analysis of rRNA database sequences.

Not withstanding few of the limitations of TRFLP, such as overestimation of diversity by 'pseudo T-RFs' and non-specific action of restriction enzymes, it has become a valuable profiling technique for rapid and sensitive estimation of temporal and spatial variations in microbial communities

viii) Multiplex T-RFLP: One of the advances in T-RFLP analysis is the demonstration of multiplex T-RFLP (M-TRFLP) which is useful in simultaneous profiling of multiple taxonomic groups of micro-organisms (two or four different taxa) within an ecosystem. M-TRFLP analysis was validated using multiple primer sets targeted to bacteria, archaea and fungi in the same PCR reaction to study different microbial taxa in an ecosystem. This multiplex molecular profiling is useful in identifying bio-indicators of pollution, environmental health and to study how differently various microbial taxa respond to environmental stress.

ix) Real-time quantitative PCR (qPCR): Real-time quantitative PCR (also referred to as qPCR) has emerged as a promising tool for rapid, reproducible and accurate estimations of microbial community dynamics or monitoring their catabolic activity during active bioremediation processes. The principle of qPCR assay is based on real-time detection of a reporter molecule whose fluorescence increases as the PCR product accumulates during each amplification cycle.The fluorescence chemistry in the qPCR reactions are either based on hybridization probes (TaqMan-molecular beacons with (FRET) fluorescence resonance energy transfer) or utilize double stranded DNA intercalating dyes SYBR Green along with carboxy-X-rhodamine (ROX) as a passive reference dye. However, to design probes and primers, the "signature sequences" unique to a particular mirco-organism or a catabolic gene need to be determined by comparison with database sequences using alignment tools. In each qPCR assay a known concentration of standard DNA (usually a linearized plasmid or genomic DNA) is used to prepare standard curves for quantification of unknown target microbial genes.

The initial amount of target DNA is inversely proportional to the cycle threshold (CT) value which can be defined as the amplification cycle when the signal of fluorescence in the assay is statistically significant above the baseline level of fluorescence. Based on cycle threshold (CT) values, the relative abundance of specific group of microorganisms in the total microbial community DNA can be quantified by targeting either taxon/species/phylum specific rRNA genes or any other catabolite biomarker genes. A disadvantage of qPCR is optimization of amplification efficiencies and PCR biases in each run for accurate quantification.

A naphthalene hydroxylating dioxygenase (nahAc) genes of naphthalene-degrading Proteobacteria within soil samples recovered from large-scale remediation processes was validated using real-time PCR assay. Copy numbers of a functional gene that encodes the alpha subunit of the PAH-ring hydroxylating dioxygenases (PAH-RHDa) within bacterial populations capable of degrading PAHs by aerobic metabolism in soil and sediment samples. Desai and his working group (2009) determined abundance of active bacterial populations of an enriched bacterial consortium-AIE2 during the steady-state condition within continuous bioreactors treating Cr(VI) and azo dye mixtures by calculating 16S rRNA gene copy numbers using qPCR assays.

15.2. MOLECULAR TECHNIQUES FOR *IN SITU* MONITORING OF MICROBIAL COMMUNITIES, BIOREMEDIATION PROCESSES AND ENVIRONMENTAL POLLUTION

Many molecular techniques have been applied for *in situ* monitoring of microbial communities involved in on-site bioremediation processes, i) such as fluorescence in situ hybridization (FISH) and ii) stable isotope probing of nucleic acids (SIP). Likewise, molecular microbial biosensors or bioreporters are developed for in situ detection of environmental pollution. These molecular *in situ* techniques are used in linking the structure and function relationships of microbial communities and for biological sensing of pollution at anthropogen contaminated sites.

15.2.1. Fluorescence *in situ* hybridization (FISH)

The FISH technique is based on selective hybridization of rRNA targeted fluorescent dye-labeled oligonucleotide probes to the ribosomes of permeabilized microbial cells prefixed on membrane filters or glass slides. The resultant microbial cells stained by the complementary rRNA-targeted probes can be visualized or counted using epifluorescence microscopy, confocal laser scanning microscopy (CLSM), or flow cytometry techniques. Multiple group-specific rRNA probes targeting prokaryotic and eukaryotic microbial taxa can be used in a FISH experiment for simultaneous phylogenetic classification as well as quantification of physiologically active microbial populations in an environmental sample. In the FISH approach it is assumed that actively growing microbes have many ribosomes and should theoretically yield brighter fluorescence signals due to higher rRNA-targeted probe hybridizations. This assumption does hold true for many microbial cells that are smaller in size, slow growing or starving or containing low cellular rRNA content Eg. *Dehalococcoides*

To over-come these limitations and improve the sensitivity of conventional FISH techniques, two new combinative approaches have been developed, namely (CARDFISH) catalyzed reporter deposition-fluorescence in situ hybridization and (FISH-MAR) fluorescence in situ hybridization-microautoradiography. Microautoradiography (MAR) is a process that relies on uptake of radioactive substrates by growing cells; the radioactivity incorporated into these cells is then visualized using radiation-sensitive photographic emulsions and microscopy.

Therefore, coupling of FISH with microautoradiography (FISH-MAR) facilitates both phylogenetic as well as functional identification of substrate-active cells within complex microbial communities.

i) **The FISH-MAR:** The FISH-MAR technique involves short incubation of the environmental sample with radio-actively labeled substrate, followed by identification of microbial populations using FISH and in-parallel processing of identified microbial cells using radiation-sensitive photographic silver emulsions. Consequently, the silver particles deposited around the actively growing cells are visualized under transmission electron microscopy (TEM) to determine whether the microbe identified using rRNA-targeted probes was functionally or metabolically active in consuming the radio-labeled substrate offered at the time of incubation.

   Applications of FISH-MAR: FISH-MAR technique is commonly used to identify key biodegradative microbial phylotypes within activated sludge systems owing to easy availability of sludge biomass for fixation, staining and hybridization experiments. In a study using FISH-MAR, it was observed that [^{14}C] glucose-degrading microbial communities were dominant in terms of abundance and diversity as compared to fatty acids-[^{14}C] propionate-[^{14}C] butyrate-utilizing microbial communities. Moreover, despite the dominance of Betaproteobacteria in the community structures, members of *Chloroflexi, Smithella, Syntrophomonas* and *Methanosaeta* groups were more capable of utilizing radio-labeled sugars and fattyacids.

ii) **Catalyzed reporter deposition (CARD)-FISH:** It is an *in situ* signal amplification technique based on the deposition of fluorochrome-labeled reporter molecules (such as, tyramine) due to catalytic activity of horseradish peroxidase (HRP) enzyme coupled with oligonucleotide probes.This leads to an enhanced fluorescence signal intensity (>12 folds) and increases the sensitivity of microbial detection using FISH probes.

   Applications of CARD-FISH: The microbial composition and structure of granular sludge biofilms in chemolithotrophic denitrifying upflow anaerobic sludge bed (UASB) bioreactors treating waste waters were studied using CARD-FISH technique It was identified that *Thiobacillus denitrificans* is the dominant microbial populations during changes in bioreactor from methanogenic to chemolithotrophic denitrifying conditions.The genus *Dehalococcoides* have a capacity to reductively dechlorinate chlorinated organic pollutants. CARD-FISH technique detects vinyl chloride (VC)-reductively-dechlorinating microbial cells in enrichment cultures using *Dehalococcoides* specific probes.

15.2.2. Nucleic acids-based stable isotope probing (SIP)

Nucleic acids-based stable isotope probing (SIP) is a novel approach which directly links the microbial community structure with its function without the need for cultivation of individual micro-organisms. In principle, SIP technique consists of providing heavy isotope-labeled substrates (e.g. ^{13}C-labeled substrates) to microbial communities and separation of the total cellular pool of nucleic acids within these microbial communities by isopycnic density gradient ultra-centrifugation. The total extracted nucleic acids will form two different centrifugal zones, one with ^{13}C-labeled (high buoyant density) and the other with C^{12}-containing nucleic acid fragments (low buoyant density). Functional microbial communities that utilized the heavy isotope-labeled substrates can be identified from the resultant ^{13}C-labeled nucleic acid fragments using molecular techniques discussed above. Natural C^{12}-containing nucleic acid fragments are usually used as negative control in the SIP experiments to differentiate between active (^{13}C-labeled) and inactive (C^{12}-containing) microbial populations.

Applications of SIP technique in bioremediation: SIP technique has been applied using wide variety of xenobiotic compounds to delineate the active microbial populations that utilize these compounds as substrates in cellular metabolism. The SIP-approach in complementation with molecular fingerprinting or sequencing method can identify potential degraders of xenobiotics; however, the actual biodegradative pathways cannot be identified using this technique.

Singleton *et al.* (2005) supplied ^{13}C-napthalene, ^{13}C-salicylate and ^{13}C-phenanthrene into PAH contaminated soils of a bioreactor to identify enriched bacterial degraders of PAHs within the bioreactor community DNA. In this DNA-SIP experiment, analysis of distinct 16S rDNA-based taxa within ^{13}C-labeled community nucleic acids using DGGE fingerprinting technique identified *Pseudomonas* sp., *Ralstonia* sp. as degraders of naphthalene and salicylate and *Acidovorax* sp. as degraders of phenanthrene hydrocarbons.

15.3. ENVIRONMENTAL METAGENOMICS IN BIOREMEDIATION

The concept of environmental genomics is based on simultaneous analysis of genes within environmental microbes (genomics) or analysis of collective microbial genomes in an environmental sample (metagenomics). Availability of whole genome sequences from several environmental micro-organisms pertinent to bioremediation is useful to determine the gene pool of enzymes involved in degradation of anthropogenic pollutants. Metagenomics surmounts the major limitations of cultivation dependent studies, as it involves extraction of nucleic acids directly from environmental samples which theoretically embodies the entire set of microbial community genomes present in a given ecosystem. Over the past few years, metagenomics based methods have been useful to determine novel gene families and or microbes involved in bioremediation of xenobiotics.

Recently, DNA microarrays is being applied to microbial ecological research in monitoring microbial communities and efficacy of several bioremediation processes.

15.3.1. Metagenomic libraries and pyrosequencing

Metagenomic libraries are constructed by direct cloning of DNA fragments extracted from an environmental sample in a suitable vector (e.g. plasmid, phage, fosmid, cosmid or bacterial artificial chromosomes BAC), which is then transformed into a suitable host strain. Cloned DNA fragments are then analyzed using either sequence-based or function-based screening procedures. These have been already discussed in previous chapters.

Many genes involved in bioremediation have been reported by constructing metagenomic libraries. Few examples are mentioned below:

Martin and his coworkers (2006) constructed metagenomic libraries to decipher ecological and metabolic functions of microbial communities involved in enhanced biological phosphate removal (EPBR) systems. The complete genome and the associated phosphate accumulation genes within an uncultured, yet dominant poly-phosphate accumulating micro-organism (POA) known as *Candidatus, Accumulibacter phosphatis* were detected which is an important finding in EPBR systems. Fosmid libraries from metagenomic DNA fragements recovered from sludge sample revealed extradiol dioxygenases (EDOs) genes using catechol as a substrate,

Metagenomics has now become a widespread approach to discover novel biocatalysts or gene products involved in biodegradation of anthropogenic compounds. However, the scope of constructing metagenomic libraries from environments having lower microbial abundance is very limited.These limitations have been over-come by application of whole community genome amplification (WCGA) based on principle of multiple displacement amplification (MDA), improving the accessibility and efficacy of metagenomic gene discoveries from low biomass environments. In this technique, all the metagenomic DNA is evenly amplified which ensures the representativeness of community microbial genomes yet another innovative technical breakthrough in the field of metagenomics is the massively parallel pyrosequencing also known as metagenomic pyrosequencing.

Many pyrosequencing based chemistries and instruments are now commercially available, such as the Genome Sequencers from Roche/454 Life Sciences [GS-20 or GS-FLX;], In addition to the massive parallelization, the 454 technology does not require cloning from metagenomic DNA, thus eliminating many of the problems that are associated with this step of metagenomics. The only limitation of pyrosequencing is the short read lengths of approximately 250–400 bp that provide poor phylogenetic information as compared to full length 16S rRNA gene sequences (1500 bp). However, these limitations can be over-come by using error-correcting barcoded primers or by accurate taxonomic assignments of 16S rRNA sequence reads error-correcting DNA barcodes that allow massively parallel pyrosequencing in a single run. In this approach, rRNA community amplicons derived from different samples were mixed and after sequencing, their data was separated according to their barcodes. In light of this revolution in massively parallel pyrosequencing of microbial rRNA genes, the recent edition of RDP release 10.10 provides an online Pyrosequencing Pipeline (http://pyro.cme.msu.edu/) that classifies, aligns and converts the pyrosequencing data to formats suitable for common ecological and statistical packages.

15.3.2. DNA microarrays

DNA microarrays are glass chips fabricated with different types of probes (pre-synthesized PCR products, cDNA, oligonucleotides, and known genomic fragments). These probes are deposited or spotted on the glass surface using metal pins (contact printing) or by ink-jets (non-contact printing). In addition, high-density oligonucleotide microarrays are generated by synthesizing probes at discrete locations using photodeprotection by photomasks/digital mirrors/chemical deprotection and ink-jet printing. The probes ranging from sizes of approximately 25–1000 bps can be used to generate homogeneous microarrays (probes from a single genomic source) or heterogeneous microarrays (probes from different genomic sources).

In principle, DNA microarray technique is based on hybridization of the target DNA molecules (single cell genomes or community genomes) to the array probes detected by measuring change in fluorescence signals (probe or target DNA can be tagged with several fluorescent dyes such as Cy3 or Cy5). The fluorescence signals from each of the probe-target hybridization spots are designated for quantification using mean signal intensity of each signal spot relative to its local background by signal-to-noise ratio (SNR) and measured using commercially available image analysis softwares. DNA Microarrays are amenable for rapid, sensitive and quantitative as well as simultaneous monitoring of several microbial populations within complex ecosystems.

Based on the probes utilized in the fabrication of an array or depending on their applications, microbial ecological microarrays can be classified into several different types. Of these different types of ecological microarrays; phylogenetic oligonucleotide arrays (POA), functional gene arrays (FGA) and whole-genome arrays (WGA) are most frequently employed in bioremediation studies.

15.3.2.1. *Phylochips in bioremediation*

Phylogenetic oligonucleotide arrays (POA) or PhyloChips are constructed using short stretches of known oligonucleotide sequences based on rRNA genes from different microbial phyla and are amongst the most commonly used microarrays to decipher microbial community structures in environmental samples. Due to huge numbers of rRNA gene sequences available from public databases along with easy in silico accessibility of online rRNA-targeted probe design and probe match tools, designing probes for a POA experiment becomes very convenient and user friendly. However,shorter stretches of oligonucleotide probes are less effective in resolving species-level phylogeny for some bacterial lineages.16S rRNA gene-targeted oligonucleotide microarray (RHC-PhyloChip) consisting of 79 probes for diversity analysis is successful in detection of uncultured Zoogloea, *Ferri bacterium/Dechloromonas-*, and *Sterolibacterium* related bacterial lineages from the activatedsludge samples.An advanced version of PhyloChips is the Isotope Array, which is based on incorporation of radioactivity into rRNA of microbes incubated with radioactively labeled substrates used to assess microbial community diversity and activity simultaneously using fluorescence and radioactivity detection modules.

15.3.2.2. *Functional gene arrays (FGAs)*

Functional gene arrays (FGAs) are designed to detect key metabolic enzymes or functional gene products involved in various biogeochemical and environmental processes. To construct FGAs containing large DNA fragments the probes are generally PCR amplified from environmental clones or bacterial strains, however it is a cumbersome and challenging task to obtain functional genes from various genomic sources. In this context, due to ease of construction and higher specificity in designing, use of an oligonucleotide FGA is more advantageous. In designing functional gene oligonucleotide probes it needs to be considered that the probes used in fabrication are evolutionarily conserved and have enough sequence divergence within representative protein families. Designing such oligoprobes is possible by comparison of alignment clusters of protein sequences obtained from public databases and identifying unique functional gene sequences.

He *et al.* (2007) developed a comprehensive FGA, termed as GeoChip, containing 24,243 oligonucleotide (50 mer) probes and covering 410,000 genes in 4150 functional groups involved in nitrogen, carbon, sulfur and phosphorus cycling, metal reduction and resistance, and organic contaminant degradation. The developed GeoChip was successful in tracking the dynamics of metal-reducing bacteria and functional activities of indigenous microbial communities associated with in situ bioremediation of uranium contaminated sites.

15.3.2.3. *Whole-genome arrays (WGA)*

Whole-genome arrays (WGA) are spotted as imprints of the complete genome of a particular micro-organism and used to contrast or correlate genomes of related micro-organisms. This WGA was used to compare and contrast genes involved in dehalogenation of chlorinated solvents, such as tetrachloroethene (PCE) and trichloroethene (TCE) from an enrichment culture containing unsequenced ‘*‘Dehalococcoides*” strains.

15.3.2.4. *'Laboratory-on-a-chip' (LOC)*

Recent technological developments in microlithography, micromachining and bonding techniques has enabled researchers to create small features, such as wells, channels, electrodes and filters on an enclosed silicon, glass or polymer substrates to device fully-integrated miniaturized systems called 'laboratory-on-a-chip' (LOC). LOC technology hold promises for providing automated, high-throughput and on-site solutions for performing different purifications and biochemical assays on a single array chip.

15.4. METATRANSCRIPTOMICS IN BIOREMEDIATION

Transcriptomic or metatranscriptomics tools are used to gain functional in-sights into the activities of environmental microbial communities by studying their mRNA transcriptional profiles. Recent advances in direct extraction of mRNA from archaeal, bacterial and eukaryotic microbial cells have enabled researchers to obtain the gene expression profile of the entire microbial community also known as "metatranscriptome". Extraction of this metatranscriptomes coupled with pyrosequencing or construction of cDNA microarrays provides a useful tool to monitor transcriptional activities of entire microbial communities. Using metatranscriptomic approach information on both structure and function of soil microbial communities can be simultaneously obtained. Transcriptomics approach involves extraction of total community RNA and reverse transcription into cDNA and pyrosequencing of resultant cDNA.

15.5. METAPROTEOMICS IN BIOREMEDIATION

Recently, traditional proteomic approaches are being applied to detect protein expression profiles directly from mixed microbial communities of an environmental sample reflecting their actual functional activities in a given ecosystem. These "community proteomics" or "metaproteomics" approaches have significantly advanced because of technological developments in 2-DE coupled with mass spectrometry (MS) techniques, along with upgradations in protein sequence and structure databases. Matrix assisted laser desorption ionization time-of-flight (MALDI-ToF) mass spectrometry is the most commonly used MS technique for identifying proteins of interest excised from 2-DE gel spots, based on the principle of peptide mass fingerprinting. Likewise, liquid chromatography linked to MS via electrospray ionization source (LC-ESI-MS) is also used to separate and identify peptide fragments from different microorganisms. Identification of peptide mass fingerprints generated by MALDI-ToF-MS or LC-ESI-MS in a metaproteomic analysis is achieved by their comparisons with previously known peptide fingerprints from the protein data banks.

A functional metaproteomic analysis was conducted on microbial communities within 2,4-dichlorophenoxy acetic acid (2,4-D) contaminated soil and chlorobenzene-contaminated ground waters of an aquifer by Benndorf *et al.* (2007). In this study, metaproteomes were separated from humic soil matrices by alkaline phenol treatment and subjected to two-dimensional-electrophoresis (2-DE) and the resultant spots of individual proteins were excised and identified by LC-ESI-MS. In this metaproteomic analysis, 2,4-dichlorophenoxy acetate dioxygenase, chlorocatechol dioxygenases, molecular chaperons and transcription factors were identified within the chlorophenoxy acid-degrading and chlorobenze-degrading microbial communities.

15.6. MICROBIAL METABOLOMICS AND FLUXOMICS IN BIOREMEDIATION

Beyond genomics, transcriptomics and proteomics, the research forefronts are now expanding towards the global analysis of the entire repertoire of cellular metabolites within a microbial cell, this newly introduced approach is known as "metabolomics". A microbial cell releases a number of low molecular weight primary and secondary metabolites in response to an environmental challenge or stress. Metabolomics approaches aim at quantifying functional roles of these metabolites in the microbial cells using separation and analytical techniques. The microbial metabolomics toolbox encompasses several methodologies like metabolic fingerprinting and foot printing, metabolite profiling and target analysis for identification and quantification of a wide range of cellular metabolites using several analytical techniques such as, nuclear magnetic resonance (NMR), direct injection mass spectrometry (DIMS), Fourier transform-infrared (FT-IR) spectroscopy, high performance liquid chromatography (HPLC), gas chromatography (GC) and capillary electrophoresis (CE) based mass spectrometry. Several studies have recently applied microbial metabolome analysis to study biodegradation of anthropogenic pollutants. For instance, comparative metabolome analysis of Sinorhizobium sp. C4 during the degradation of phenanthrene indicate several intermediates like trehalose, branched-chain amino acids and intermediates of tricarboxylic acid cycle and glycolysis.

It is well established fact by now, that metabolism is a regulated mechanism at genomic, transcriptional, and posttranslational levels. Hence, a relatively new concept focusing on real-time flux analysis of cellular molecules/ metabolites within a cell over a time period known as "fluxomics" has been proposed by

Wiechert *et al.*, 2007. Fluxomics analysis Shewanella sp. revealed co-metabolic pathways for bioremediation of toxic metals, radionuclides, and halogenated organic compounds. In this study, the metabolic flux analysis of Shewanella sp. was depicted using GC-MS and statistical, biochemical and genetic algorithms.

15.7. APPLICATION OF METAGENOMICS IN BIOREMEDIATION – SOME SUCCESSFUL EVENTS

15.7.1. Biological characterization of ammonia-oxidising bacteria within a domestic effluent treatment plant treating industrial effluents

15.7.1.1. *Nitrification and harmful effect of ammonia*

Nitrification is an important design and process consideration in wastewater treatment systems for the removal of ammonia. Nitrification is a microbiologically-mediated two-stage process. The first stage results in the oxidation of ammonia to nitrite via hydroxylamine. The second stage oxidises nitrite to nitrate. Ammonia has an acute toxicity to fish and other aquatic species, and can lead to eutrophication of environmental waters. Due to the sensitivity and slow-growing nature of nitrifying bacteria, nitrification is readily inhibited by a variety of toxic components that may be present in wastewater. The mechanism and level at which inhibition is elicited is not understood and is consequently difficult to measure, predict and control.

15.7.1.2. *Demerits of current nitrification assessment methods*

Current methods to assess nitrification inhibition are relatively time consuming and insensitive. They also rely on measurements based on foreign, unrepresentative microorganisms whose physiological properties, and susceptibility to inhibition may differ from that of the indigenous nitrifying bacteria. The development of a rapid realistic test to determine nitrification inhibition would be invaluable in the optimization of nitrification in wastewater treatment.

15.7.1.3. *Molecular tools for nitrification inhibition studies*

A useful approach to assess nitrification inhibition is to monitor the actual response of the nitrifying population present in wastewater. Ribosomal ribonucleic acid (rRNA)-based molecular techniques provide the tools to achieve this. Molecular studies allow the direct determination of the effects of inhibiting material on indigenous bacteria. Furthermore these techniques have the potential to quantify the level of inhibition and determine whether specific inhibitors differentially affect particular members of the bacterial population. Molecular techniques such as denaturing gradient gel electrophoresis (DGGE) and differential "staining" of specific bacterial cells using fluorescence *in situ* hybridization (FISH) can be used to characterize the ecology and population dynamics of the nitrifying bacteria. The autotrophic AOB responsible for the initial oxidation of ammonia to nitrite are the primary focus. The reason for this is that nitrification is driven by this initial step and ammonia oxidation is thought to be the most sensitive step, as nitrite rarely accumulates.

Case study I

In one of the research project in US, biomass samples were obtained from the activated sludge pretreatment plant, primary and secondary trickling filter beds, and the Biopur biological aerated filter. Total nucleic acid was extracted and purified from each sample and 16S ribosomal deoxyribonucleic acid (16S rDNA) sequences amplified by polymerase chain reaction (PCR) using primers that were specific for AOB. PCR products were analysed by denaturing gradient gel electrophoresis (DGGE), cloned, sequenced and compared to 16S rRNA databases. Fluorescent oligonucleotide probes were then designed to allow the quantification of the most abundant AOB present in the biomass using FISH (Figure 15.2)

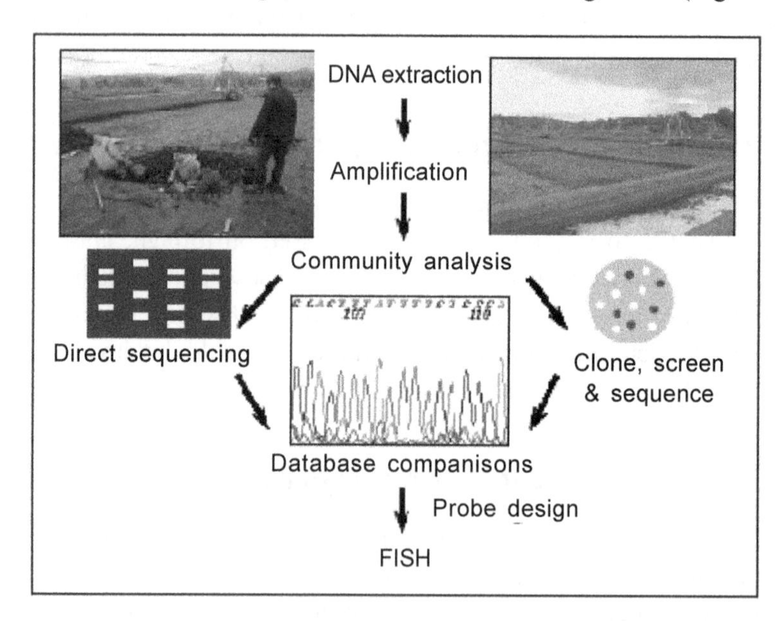

Fig. 15.2. Detection of ammonia oxidizing bacteria in a domestic effluent treatment plant

PCR using AOB specific primers and subsequent community analysis using DGGE highlighted a number of bands. Each band corresponds to the presence of at least one distinct bacterium. A wide diversity of AOB were found throughout the treatment plant with the greatest diversity in the secondary filter bed. Differences and similarities could be observed between the reactor configurations. One dominant band was observed within all reactor configurations.

The principal bacterial component of NitrotoxTM, a commercial preparation of nitrifying bacteria used in inhibition studies, was identified as *Nitrosomonas eutropha*. This was not detected in any of the samples studied. This is a particular concern as toxicity-based consents are determined using this commercial bacterial preparation by the water authority accepting the industrial effluent. Cloning and sequencing of DNA sequences from the reactors confirmed that all the bacterial sequences amplified by PCR were members of the *beta*-proteobacteria. The majority of the clones sequenced also clustered with the *Nitrosomonas* subgroup of AOB with the dominant AOB present in all bioreactors being closely related to *Nitrosococcus (Nitrosomonas) mobilis*. This is a halotolerant AOB that was first isolated from seawater. None of the clones obtained clustered *with Nitrosomonas europaea*, the AOB for which there are physiological data due to its ease of culture within the laboratory. FISH studies using probes designed specifically against *Nitrosococcus mobilis* were used to detect the relative abundance of the target AOB in all bioreactor samples.

AOB are present throughout the effluent treatment plant, even in reactors where nitrification inhibition was thought to exist. The greatest diversity of AOB populations was found in the secondary filter bed. *Nitrosococcus mobilis* is the dominant AOB throughout the treatment plant. This organism is known to be halotolerant. The AOB present in Nitrotox^{TM} is not indigenous to the treatment plant. This raises some serious questions regarding its use in toxicity assessment and bioaugmentation.

15.7.2. Exploiting metagenomics to develop a terrestrial biomarker for heavy metal contamination

Understanding the risks associated with environmental contamination with specific heavy metals such as copper, cadmium, lead and mercury has been a subject of investigation throughout the last century. All metals, whether biologically essential or otherwise, pose two fundamental problems: they are more-or-less toxic above certain threshold concentrations and they are not biodegradable. In the field of human health, metals continue to cause major diseases, from copper causing

Indian Childhood liver cirrhosis (Sethi *et al.*, 1993) to exposure to cadmium in drinking water or food resulting in Itai-itai disease (Nogawa., *et al.* 1996). In addition to clinical considerations/implications, transfer and bioamplification of metal ions within food chains has been shown to have significant impacts on ecosystems. Therefore, when evaluating environmental emissions, assessing risk from contaminated land or legacy sites, and determining chronic exposures due to changes in waste management practices, both through landfill disposal and sewage sludge processing the heavy metal dimension remains a primary consideration. However, assessing the risk posed by heavy metals is complicated by a suite of mitigating factors that influence metal bioavailability and therefore the relative toxicity.

Therefore, the emerging questions are: could metagenomics approach supply an objective marker for metal contaminated land supplying not only a measure of environmental risk but also linking individual responses to the population level impacts? In other words, could functional parameters of free-living organisms provide quantitative measures of the cocktail of poorly-understood interactions leading to metal toxicity? Preliminary work has focused on identifying a marker for cadmium toxicity using a resistant population of the established terrestrial sentinel earthworm, *L. rubellus*, inhabiting a Roman mine site. Initially, classical protein biochemistry was used to identify a cellular constituent responsible for binding the bulk of the cadmium constituting approximately 1% of the body burden in this population. When purified, this protein displays many features, spectral and amino acid composition, associated with a family of well characterised, small, cysteine rich, metal binding proteins, the metallothioneins (MTs, Stürzenbaum *et al.*, 1998).

15.7.2.1. *Metal binding motifs the metallothioneins*

A molecular approach exploited specific metal binding motifs associated with chelation of cadmium. This allowed a genetic fingerprint to be generated of genes containing this motif from control populations and compared with that exhibited by worms exposed to cadmium. Isolation and characterization of the major genetic element of this fingerprint, which was specific to the cadmium-exposed organisms, revealed it to encode earthworm MTs. Using this fragment from the genetic fingerprint, two MT isoforms were isolated from earthworms. Subsequently it was confirmed that the gene encoded by MT2 represented the protein isolated from the earthworms within the initial phase of the study (Morgan *et al.*, 2003). Furthermore, by expressing both isoforms within bacterial cells, substantial material could be generated which facilitated the production of a polyclonal antisera, that had been used to demonstrate the involvement of MT2 in the uptake, sequestration and excretion of cadmium within the earthworm whilst MT1 displays no direct involvement in cadmium detoxification (Stürzenbaum *et al.*, 2001). Having demonstrated that MT2 is mechanistically linked to cadmium sequestration, it was important to validate it as a quantitative marker of Cd exposure and to link it to lifecycle alterations caused by metal toxicity. The rationale behind the latter approach is that genetic changes must precede and, therefore, predict whole organism and population level (i.e. ecologically relevant) functional levels. This demonstrated a significant correlation between MT2 expression and the intrinsic rate of population increase. However, although a mechanistic link has been unequivocally demonstrated between body cadmium levels and MT2 expression, as yet no direct mechanistic evidence links MT expression with reproduction. However, precedence for this link exists in fish, where MT interacts with the systems responsible for the generation of the egg yolk protein vitellogenin.

15.8. TOXICOGENOMIC PROFILING OF BIOREACTIVE PARTICLES WITHIN DIESEL EXHAUST

Epidemiological studies conducted first in the USA and later in the UK, suggested that a relationship exists between increasing cardio respiratory hospital admission, morbidity and mortality rates and increase in the small particulate matter, originating from fumes such as diesel exhausts. PM10s are particulates with an aerodynamic diameter of less than 10 μm, they are complex mixtures of natural materials, metals, carbonaceous components, soluble ionic species and organic micro-pollutants. In urban environments, diesel exhaust particles (DEP) form a large constituent (20-80%) of the airborne PM10s arising from vehicular activities. The exact biological mechanism by which these molecules elicit their toxicological effect is unclear. Furthermore, the impact of these particles exert on the wider environment is poorly understood. To this end, a toxicogenomic approach was employed to profile the genes involved in the toxicological response by Reynolds and Richards, (2001). The samples were subjected to following analysis.

(a) Chemical analysis
(b) Cellular studies (mostly to look at oxidative metabolism)
(c) *In vitro* toxicology studies with lung epithelial target cells (human and animal primary cells)
(d) *In vivo* lung toxicology following instillation (direct deposition via the trachea under anaesthesia).

The rationale was to look for genetic markers or candidate genes to monitor the toxic response. For this a commercial macroarray, representing genes involved in stress, cell signalling, xenobiotic metabolism, DNA repair/cell cycle, inflammation etc was employed. This stress macroarray includes 207 rat cDNAs, double-stopped on a positively charged nylon membrane. In negative controls, housekeeping genes are included to ensure hybridization specificity and as positive controls. Identical membranes were hybridised to radiolabel targets generated from the genes transcribed within control and PM10 exposed animals. After high stringency washing, the quantity of hybridisation was analysed using phosphorimaging. This allowed the relative level of each transcript to be determined during a time course of exposure to different environmental PM10 samples. Of particular interest were the water soluble component of PM and specifically its bioavailable transition metals. It was hypothesised that these are the problem agents and by being bioavailable are crossing the lung barrier and at first pass hitting the heart. This might explain why people exposed to air pollution die of cardiac problems. The studies are being extended to a range of body tissues and are expected, within a few years, to identify the most important transition metal in the mixture (believed to be zinc). More importantly, this technique is revealing how the heart talks to the lung and how relevant drugs may compromise lung defences and modulate the affects on the heart.

15.9. ANAEROBIC AMMONIUM OXIDIZING BACTERIA AND *NITROSPIRA*-LIKE BACTERIA

Application of the full-cycle rRNA approach has revealed that in many ecosystems previously not recognized bacteria catalyze key steps of the biogeochemical nitrogen cycle. For example, uncultured members of the phylum *Nitrospirae* are important nitrite oxidizers in many natural habitats and in wastewater treatment plants. These novel nitrite oxidizers are members of a unique bacterial phylum, which harbors also other highly interesting bacteria like the thermophilic sulfate reducers of the genus *Thermodesulfovibrio* and the acidophilic iron oxidizers of the genus *Leptospirillum*. In addition, anaerobic ammonium oxidation (Anammox) is mediated by novel, uncultured, deep branching members of the Planctomycetales, which are of significant biotechnological interest.

The goal of this project is to sequence large genome fragments or even the entire genome of a Nitrospira-like nitrite oxidizer and of an Anammox organism. For this purpose, high molecular weight DNA was extracted from enrichments of the respective organisms, purified by pulsed field gel electrophoresis, and partially digested. DNA fragments were cloned into bacterial artificial chromosomes (BAC). In total, approx. 9,000 BAC clones from the enrichment of the Anammox and

4,500 BAC clones from the Nitrospira-enrichment were obtained (insert size ranging from 30 to 200 kb) and both ends of each BAC insert were sequenced. In addition, encompassing shotgun libraries containing more than 100,000 clones were constructed and sequenced for both organisms. Large contigs were assembled. Subsequently, two approaches were applied to identify contigs of the target organisms. BACs carrying the 16S rRNA gene of the target organism were identified by membrane hybridization with suitable gene probes. In addition, peptide sequences of a key enzyme and a cytochrome of Anammox (determined by separate proteome analysis of the enrichment) were used to identify Anammox contigs by sequence homology searches. Genomic data will provide new insight into the biology of the elusive *Nitrospira*-like bacteria and Anammox organisms.

15.10. MECHANISTIC INSIGHTS INTO TOXICITY OF AND RESISTANCE TO THE HERBICIDE 2, 4-D.

Genomic information and tools are beginning to be used to increase our understanding of how organisms of all types interact with their environment. The study of the expression of all genes, at the genome, transcriptome, proteome and metabolome level, in response to exposure to a toxicant, is known as toxicogenomics. Metagenomics has enhanced the development of fundamental knowledge on the mechanisms behind the toxicity of and resistance to the herbicide 2,4-dichlorophenoxyacetic acid (2,4-D). Although 2,4-D is one of the most successfully and widely used herbicides, its intensive use has led to the emergence of resistant weeds and might give rise to several toxicological problems when present in concentrations above those recommended.

15.11. CONCLUSION

Microorganisms possess powerful but limited capabilities to respond to changing environmental conditions. We have some understanding of phenotypic responses to environmental conditions and the control mechanisms associated with relevant, specific genes. A more integrated understanding of the complexity of phenotypic and genetic responses will be gained from research that addresses these issues on a genome-wide scale, but relatively few studies at this scale have been conducted. Additionally, while populations and species possess an even greater collective diversity of responses than do individuals, applying metagenomic approaches to understand the diversity and functions of population through ecosystem scales remains a major challenge. Metagenomic approaches that connect these scales of analysis offer the opportunity to gain novel insights into environmental issues and ecological responses to environmental change.

Mini Quiz

1. What is single strand conformation polymorphism?
2. What does a CT value infer in qPCR?
3. What is the principle of FISH? Mention the two innovative FISH methods?
4. Write the striking features of functional gene arrays?
5. Explain 'laboratory-on-a chip'?

CHAPTER - 16

METAGENOMICS IN WHITE BIOTECHNOLOGY

Different industries have different motivations to probe the enormous resource that is uncultivated microbial diversity. Currently, there is a global political drive to promote white (industrial) biotechnology as a central feature of the sustainable economic future of modern industrialized societies. This requires the development of novel enzymes, processes, products and applications. Metagenomics promises to provide new molecules with diverse functions, but ultimately, expression systems are required for any new enzymes and bioactive molecules to become an economic success. This chapter highlights industrial efforts and achievements in metagenomics.

16.1. BIOTECHNOGICAL PROSPECTS FROM METAGENOMICS

The biggest reservoir of microbial diversity remains uncultured and untapped within natural and human engineered ecosystems. Microorganisms in natural environments contain genes that encode and express biosynthetic or biodegradative pathways of interest that have never been identified using culture-dependent methods. To address this cultivation gap, functional metagenomic screens have been developed to recover active genes sourced directly from environmental samples. Some of the applications of metagenomics in biotechnological prospects are given below:

16.1.1. Natural product discovery: Enzymes

Enzymes expressed from cultured soil microorganisms have been harvested and used commercially for many decades. High-throughput screening of environmental metagenomic DNA libraries has led to the discovery of many novel enzymes that are of great use in industrial applications. Indeed, the very first metagenomic study involved the identification of cellulases from a bioreactor "zoolibrary".

16.1.1.1 *Enzymes discovered via metagenomics*

There are many examples of enzymes discovered *via* metagenomic approach, such as a multifunctional glycosyl hydrolase identified from a rumen metagenomic library (Palackal *et al.,* 2007), low pH, thermostable α-amylases discovered from deep sea and acidic soil environments (Richardson *et al.,* 2002),pectinolytic lyases from soil samples containing decaying plant material (Solbak *et al.* 2005), agarases from soil (Voget *et al.,* 2003) and lipolytic enzymes such as esterases and lipases (Rondon *et al.,* 2000; Voget *et al.,* 2003; Lee *et al.,* 2004; Ferrer *et al.,* 2005). Nitrilases were discovered from screening environmental (terrestrial and aquatic) DNA libraries using high-throughput and culture-independent methods (Robertson *et al.,* 2004). A novel β-glucosidase gene isolated by screening a metagenomic library derived from alkaline polluted soil was found to be a first member of a novel family of â-glucosidase genes (Jiang *et al.,* 2009). The discovery of a diverse set of genes that encode enzymes for cellulose and xylan hydrolysis from the resident bacterial flora of the hindgut paunch of a wood-feeding 'higher' termite (*Nasutitermes* sp.) and from moths was a result of metagenomic analysis (Brennan *et al.,* 2004; Warnecke *et al.,* 2007).

The rate of discovery is generally less than one clone with activity per 1,000 clones screened; therefore, the anticipated "hit rate" for any enzymatic activity should be considered prior to initiating metagenomic library screening. These are just a sampling of the many enzymatic activities discovered from metagenomes, providing ample evidence of the potential of this approach for the discovery of novel biocatalysts from the environment. Mining for biocatalysts from metagenomic libraries usually involves different strategies:

1) homology-driven metagenome mining based on high throughput sequencing
2) activity-based screening
    a) substrate induced gene expression (SIGEX)
    b) metabolite-regulated expression (METREX),
    c) Product induced gene expression (PIGEX)

Unlike chemical synthesis, biocatalysis does not include the use of toxic chemical reagents. The discovery of novel enzymes through these approaches is an economical and potentially environmentally responsible way to decrease the use of toxic chemicals traditionally used in many industries. This approach for enzyme discovery can help to improve the efficiency of existing techniques and also enable novel processes for the production of various chemicals that serve as precursors in the synthesis of pharmaceuticals, insecticides, fertilizers, herbicides, etc.

16.1.2. Discovery of novel Antibiotics

New antibiotics are necessary to treat microbial pathogens that are becoming increasingly resistant to available treatment. Despite the medical need, the number of newly approved drugs continues to decline. We offer an overview of the pipeline for new antibiotics at different stages, from compounds in clinical development to newly discovered chemical classes. Consistent with historical data, the majority of antibiotics under clinical development are natural products or derivatives thereof. However, many of them also represent improved variants of marketed compounds, with the consequent risk of being only partially effective against the prevailing resistance mechanisms. In the discovery arena, instead, compounds with promising activities have been obtained from microbial sources and from chemical modification of antibiotic classes other than those in clinical use. Furthermore, new natural product scaffolds have also been discovered by ingenious screening programs. After providing selected examples, we offer our view on the future of antibiotic discovery. For the past five decades, the need for new antibiotics has been met largely by semisynthetic tailoring of natural product scaffolds discovered in the middle of the 20th century. More recently, however, advances in technology have sparked a resurgence in the discovery of natural product antibiotics from bacterial sources. In particular, efforts have refocused on finding new antibiotics from old sources (for example, streptomycetes) and new sources (for example, other actinomycetes, cyanobacteria and uncultured bacteria). This has resulted in several newly discovered antibiotics with unique scaffolds and/or novel mechanisms of action, with the potential to form a basis for new antibiotic classes addressing bacterial targets that are currently underexploited.

As-yet-uncultured microorganisms are an untapped reservoir for the discovery of secondary metabolites such as antibiotics. The biosynthetic pathways encoding the secondary metabolites can be captured by cloning large fragments of contiguous metagenomic DNA into heterologous hosts that are easier to manipulate *in vitro*, such as *E.coli*. Many low molecular weight molecules are produced during specific growth phases such as during developmental stages or starvation and exhibit bioactive properties. A diverse class of secondary metabolites is the polyketides produced by modular enzymatic pathways with phenomenal structural heterogeneity and yet with some conserved DNA sequences that allow their identification via nucleic acid probes.

Table 16.1. Enzymes discovered by activity based screening

Environment	Target gene	Host, vector	Average size of insert (kb)	# of positives/# of screened clones	1 positive/Mb DNA screened	Assay technique
Compost	Esterase	*E.coli*, plasmid	3.2	14/21 00.0	1/4.8	Agar plate assay
	Amytlase			13/31 96.7	1/7.9	
	Phosphatase			38/31 96.7	1/2.7	
	Dioxygenase			2/31 967	1/51.1	
	Protease			0/31 967	0/103.3	
Compost (artifically contaminated with poly-lactic acid disks)	Esterase	*E.coli*, plasmid	2.5	3/40 000	1/33.3	Agar plate assay
Soil (nonspecified)	Xylanase	*E.coli*, plasmid	8.5	1/24 000	1/204	Agar plate assay
Loam soil	Oxygenase	*E.coli*, plasmid	5.5	1/65 000	1/357.5	Agar plate assay
Mining shaft, compost soil, sediments (sewage plant, sea, lake, and river)	Protease	*E.coli*, plasmid	4.0	4/389 000	1/389	Agar plate assay
Gypsy moth midgut	Production of *N*-acyl homo-serine lactons	*E.coli*, plasmid	3.3	1/1800 000	1/2640	Reporter assay
Rhizosphere soil from near side of acid mine drainage	Nickel resistance	*E.coli*, plasmid	2.5	13/726 500	1/139.7	Growth assay
Clacial ice	DNA polymerase I	*E.coli*, plasmid	4	230/230 000	1/4	Growth assay

[Table Contd.

Contd. Table]

Environment	Target gene	Host, vector	Average size of insert (kb)	# of positives/# of screened clones	1 positive/Mb DNA screened	Assay technique
Ground water from crudle-oil storage cavity	Aromatic hydrocarbon catabolic operon	*E.coli*, plasmid	7	35/152 000	1/30.4	Reporter assay
Intertial flat sediment	Lipase	*E.coli*, fosmid	N.A.[a]	1/6000		Agar plate assay
Deep-sea sediment	Lipase	*E.coli*, fosmid	32.3	1/8823	1/285	Agar plate assay
Activated sludge treating coke plant wastewater	Extradiol dioxygenase	*E.coli*, fosmid	33	91/96 000	1/34.8	Cell lysate assay
Soil (pasture)	Degradation of N-acyliho-moserine lactons	*E.coli*, fosmid	35	1/10 121	1/354.2	Reporter assay
Activated sludge treating coke plant wastewater	Bleomycine resistance gene	*E.coli*, fosmid	33	3/96 000	1/1056	Growth assay
Forest rhizosphere soils	Fungal antagonism	*E.coli*, fosmid	35	1/113 700	1/3979.5	Growth inhibition
Human fecal from healthy person and patients of Crohn's disease	Epithelial cell growth modulation	*E.coli*, fosmid	43	20/20 725 (inhinition) (30/20 725) (stimulation)	1/44.6 (inhibition) 1.29.7 (stimulation)	Cell lysate assay

[Table Contd.

Contd. Table]

Environment	Target gene	Host, vector	Average size of insert (kb)	# of positives/# of screened clones	1 positive/Mb DNA screened	Assay technique
Glacial ice	DNA Polymerase I	*E.coli*, fosmid	36	20/4000	1/7.2	Growth assay
Surface seawater	Esterase	*E.coli*, BAC	70	4/20 000	1/350	Agar plate assay
Pig fecal	Tetracycline-resistance gene	*E.coli*, BAC	15	10/9000	1/13.5	Growth assay
Rabbit cecum	Cellulase	*E.coli*, cosmid	35.1	11/32 500	1/103.7	Agar plate assay
Soil (wetland and sandbars)	4-Hydroxy-phenylpyruvate dioxygenase	*E.coli*, cosmid	40	5/30 000	1/240	Liquid-base assay
Cow rumen	Mannanase/ glucanase/ xylanase	*E.coli*, phagemid	3	1/50 000	1/150	Agar plate assay
Soil	Degradation of *N*-acylhomo-serine kactibs	*E.coli*, phagemid	4.3	3/7392	1/10.60	Reporter assay
Oil-contaminated soil	Naphthalene dioxygenase	*Pseudomonas putida*, cosmid	25	2/24 000	1/300	Growth assay

[a] N.A., Not available.

Table 16.2: Enzymes screened by sequence-based approach

Environment	Target gene	Method	Number of retrived clones	Identity to known enzyme
3-Chlorobenzoate enrichment	Benzoate 1,2 dioxygenase chlorocatechol 1,2-dioxygenase	Degenerate PCR	2	74,88%
Sediment from hot spring	Pullulanase	Degenerate PCR	1	48%
1,2-Dichloroethane enrichment	Reductive dehalogenase	Degenerate PCR	1	98%
Bioreactors treating gold-bearing concentrates	Sulfur oxygenase reductase	Degenerate PCR	2	48,53%
Deep-sea sediment	Alkane hydroxylase	Degenerate PCR	2	55,56%
4-Chlorobenzoate enrichment	4-Chlorobenzoyl-CcA dehalogenase	Degenerate PCR	2	77%
Marine sponge	Related polyketide synthesis	Degenerate PCR	3	50%
Marine sponge	Related polyketide synthesis	Degenerate PCR	5	43%
Deep-sea hydrothermal vents	Integron gene casette	Degenerate PCR	46	N.A.[a]
Tar pond	Integron gene casette	Degenerate PCR	708	N.A.[a]
Grassland soil	Nitrite reductase, nitrous oxide reductase	Probe hybridization	9	75-84%
Sequence database	Methylhalide transferase	In silico data mining	89	Average 28%

[a]N.A., not available

The adoption of heterologous hosts besides *E. coli* permits expression of cloned DNA from diverse sources. *Streptomyces* species and other *Actinobacteria* have been used as screening hosts for soil DNA libraries because of their ability to express diverse polyketide and other bioactive secondary metabolites and their relative ease of genetic manipulation. For example, the antibiotic terragine with anti-*Mycobacterium* activity was discovered via heterologous expression of metagenomic clones within a *Streptomyces lividans* host. The antimicrobial products detected in each host were distinct, supporting the contention that each heterologous host may yield a novel range of expressed metabolites from a given metagenomic library. Metagenomic clones producing long-chain N-acyltyrosine antibiotics (Brady *et al.,* 2004), antifungal agents (Chung *et al.,* 2008), and triaryl cation antibiotics turbomycin A and B (Gillespie *et al.,* 2002) were also reported.

16.1.2.1. *Antibiotics discovered by metagenomics approach*

With standard inhibition assays, a Mycobacterium inhibiting antibiotic, terragine, was discovered from a soil metagenomic clone maintained in *Streptomyces lividans* and acyl tyrosines from a clone maintained in *E. coli* (Brady *et al.,* 2001). Colored antibiotics represent a disproportionate share of those discovered because they can be identified visually. For instance, a clone noticed for its brown pigment was found to produce melanin. Methanol extract showed red and an orange pigment which on structural analysis revealed that they were triacyl cations and designated as, turbomycin A and turbomycin B (Gillespie *et al.,* 2002). Both turbomycin A and turbomycin B exhibited broad spectrum antibiotic activity against gram negative and gram positive organisms. A purple-pigmented clone (Brady *et al.,* 2001) produced violacein, previously shown to be an antibiotic made by the soil bacterium *Chromobacterium violaceum*. The sequence of the genes on the metagenomic clone diverged substantially from the *C. violaceum*. The violacein biosynthetic operon despite similar genetic organization , suggesting that the pathway on the metagenomic clone was derived from an organism other than *C. violaceum*. Two structurally related compounds, indirubin and indigo blue, were identified in a soil metagenomic DNA library based on their blue color (MacNeil, 2001).

16.1.2.2. *Antibiotics against Methicillin-resistant Staphylococcus aureus (MRSA)*

"Metagenomics has the potential to access large numbers of previously inaccessible natural antibiotics," say the researchers. Scientists are reporting use of a new technology for sifting through the world's largest remaining pool of potential antibiotics to discover two new antibiotics that work against deadly resistant microbes, including the "super bugs" known as MRSA. [Courtesy: report of *Journal of the American Chemical Society]*. Methicillin-resistant *Staphylococcus aureus* (MRSA) infections, are resistant to most known antibiotics. MRSA strikes at least 280,000 people in the U.S. alone every year, and almost 20,000 of those patients die. The typical way of discovering new antibiotics involves identifying and growing new bacteria from soil and other environmental samples in culture dishes in the laboratory. That environmental treasure-trove is the largest remaining potential source of new antibiotics. But most bacteria found in nature can't grow in the laboratory. With this "metagenomics" method, they identified two new possible antibiotics called fasamycin A and fasamycin B that killed MRSA and vancomycin-resistant *Enterococcus faecalis*, which also is becoming more resistant to known antibiotics.

16.1.3. Discovery of novel drug

One area of novel drug discovery which has been made possible by metagenomics is the investigation of symbiotic bacteria, one of the major natural sources of therapeutic products and one which has so far resisted analysis by pure culture techniques. In a PCR based screening for functionally unusual and rare type I poly-ketide synthases in metagenomes derived from beetle and sponge microsymbionts were detected. That this rare type I polyketide synthase is cloned and maintained in a heter-ologous host means that it can now be characterized revealing valuable information about polyketide enzymology and allowing generation of novel antitumour compounds.

16.1.4. Novel Industrial Biocatalysts from Metagenomics

Biocatalysis, the use of microbial cells or isolated enzymes in the production of fine chemicals, is steadily moving towards and becoming accepted as an indispensable tool in the inventory of modern synthetic chemistry. It is estimated that in 10% of the cases biocatalysis will provide an overall superior synthetic strategy over traditional organic chemistry. Application of modern screening technology to biodiversity is clearly a rewarding approach in the development of "white biotechnology field". The metagenome approach will revolutionize the field of enzyme discovery by providing genetic access to the unseen majority of microbial diversity and its enzymatic constituents (Lorenz and Eck, 2004).

Voget *et al.* (2003) cloned a soil metagenome of mixed microbial population in a cosmid and obtained cosmid libraries. Analysis of the clones from the cosmid libraries by DNA sequencing tend to identify several biocatalyst encoding genes such as amidase, cellulases, alpha amylases, 1,4 alpha glucan branching enzymes, pectate lyases, lipases and DNAases. This opened a new door for prospecting novel biocatalysts from metagenomes of any complex environment. Yun *et al.* (2004) isolated a novel amylolytic gene from a soil metagenomic library, constructed using pUC19 vector. AmyM has mixed characteristics of the alpha amylase, 4

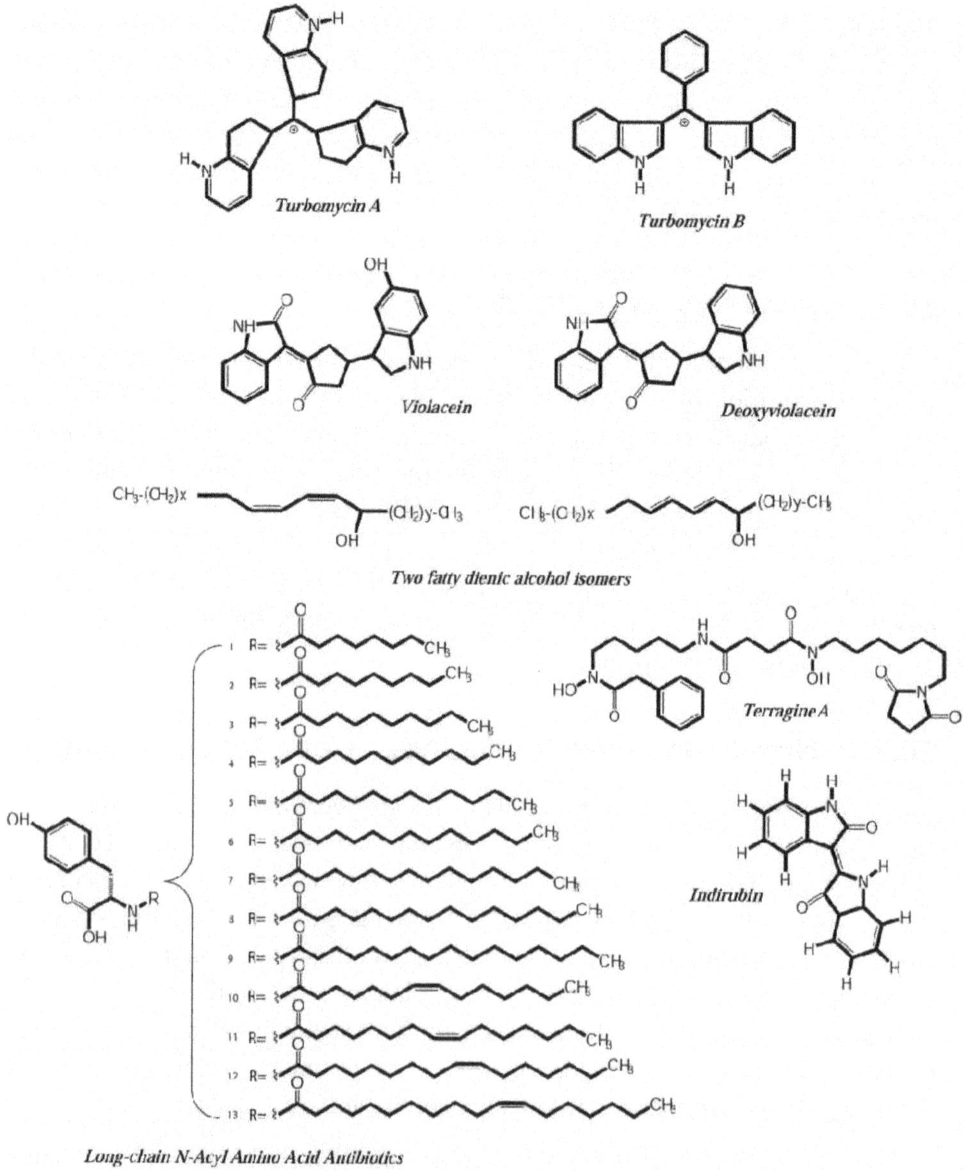

Fig. 16.1 Antibiotics discovered in metagenomic libraries

alpha glucanotransferase and neopullulanase familes. AmyM can hydrolyse starch, cyclodextrins and pullulan. Also the enzyme has a transglycosylation activity that is able to disproportionate maltooligosaccharides. These unique properties of AmyM might be due to hydrophobic differences at its catalytic site, compared to the primary structures of other related amylases.

Thermophiles are a valuable source of thermostable enzymes with properties that are often associated with stability of solvents and detergents. Rhee *et al.* (2005) isolated a gene encoding for a thermostable esterase by functional screening of metagenomes from thermal environmental samples. An amino acid sequence comparison with other esterases and lipases revealed that the enzyme should be classified as a new member of the hormone sensitive lipase family.

16.1.5. Metagenomics in biofuel industry

Cellulose, one of the most abundant sources of organic carbon on the planet, has wide-ranging industrial applications, with increasing emphasis on biofuel production. As a result,many studies have extensively focused on the identification of carbohydrate active enzymes, or CAZymes, using both culture-dependent and culture-independent methods. The CAZy database (http://www.cazy.org) currently defines 131 families of glycoside hydrolases (GHs) based onsequence and structure providing a useful resource for functional annotation of predicted GH genes (Cantarel *et al.,* 2009). Seventeen of these families are reported to have cellulase activity, classified by their ability to hydrolyse 1,4- β-d-glucosidic linkages found in cellulose, lichenan, and cereal β-d-glucans.

The current production of cellulosic ethanol from non-feedstock crops typically utilizes enzymatic hydrolysis steps to break cellulose into its constituent sugars prior to fermentation The current high cost of versatile industrial enzymes is a limiting factor in this production necessitating the discovery or development of new enzymes that may show more desirable attributes conducive to current cellulosic ethanol pipelines, such as improved acid and temperature stability. While the discovery of many different enzyme classes has been reported, cellulases have been among the most sought after genes from a biotechnological perspective. Functional metagenomic screens to identify novel cellulases have been conducted on environmental samples from soils (Nacke *et al.,* 2012), gutmicrobiomes (Pope *et al.,* 2010) and a bio-gas plant. A fosmid library containing 6144 clones sourced from a mining bioremediation system was screened for cellulase activity using 2,4-dinitrophenyl â-D cellobioside, a previously proven cellulose model substrate. Fifteen active clones were recovered and fully sequenced revealing 9 unique clones with the ability to hydrolyse 1,4- β-d-glucosidic linkages. Transposon mutagenesis identified genes belonging to glycoside hydrolase (GH) for mediating this activity.

The finding of novel cellulase family is a positive signal for the growth of biofuel industries. These cellulases can be applied widely in biotechnology firms for biooconversion of lignocellulosic biomass from agricultural residues.

16.2. FUTURE PROSPECTS

The functional metagenomics approach in bioprospecting suffers a a common problem because of the insufficient, biased expression in *E. coli.* Unlike the overexpression of a specific gene, the unbiased (or less biased) expression of foreign genes at sufficient levels requires a totally different approach. Ribosome engineering, as well as targeted engineering of factors related to transcription/translation will provide possible solutions. As for the sequence-based approach, taking advantage of progress in synthetic biology, we will be able to take a more radical approach to retrieve all possible candidate genes from the database, synthesize the genes, and test the activities. A vast and increasing volume of uncharacterized proteins exists in the public sequence databases that are rich, unexplored genetic resources, along with genes still unexplored in natural environments. Not a simple BLAST search, but rather structure-guided functional prediction, will be a key technology for obtaining novel products.

Mini Quiz

1. What is activity based on screening?
2. Mention few antibiotics discovered by metagenomics approach?
3. Name two possible antibiotics against MRSA?
4. Briefly explain novel industrial biocatalysts unraveled *via* metagenomics?

CHAPTER - 17

METAGENOMICS IN AGRICULTURE

Metagenomics refers to the genomic assemblages of microorganisms isolated directly from their environment, without the need for prior culturing under laboratory conditions. Thus the term "culture independent" often accompanies descriptions of metagenomic techniques. In more recent times, metagenomics entails massively parallel sequencing of microbial metagenomes, but functional analyses are still performed using clones. The massive diversity of bacterial and fungal genomes in soil provides a potentially vast pool of genes that code for the production of bioactive compounds. Metagenomics find its application in the field of Agriculture. Allele mining of metagenome will derive beneficial genes to alleviate both biotic and abiotic stresses. Two approaches are developed using metagenomic techniques: (1) Isolation of novel compounds produced by vector–hosts expressing the biosynthesis genes (2) Identification of resistance genes in vector–hosts exposed to high levels of a particular compound. Both approaches consist of different functional screening assays, as discussed in previous chapters, but the initial construction of the metagenomic libraries (clones containing fragments of the metagenomes) are identical. Adoption of metagenomic approaches in pest, disease, weed and stress management research can lead to novel strategies in cropping and landscape systems.

17.1. METAGENOMICS IN BIOCONTROL

The rhizosphere region of crop plants represents interesting bio-resources; their intense interaction with microorganisms is only partly understood. Both the rhizosphere and rhizoplane regions are colonized at high abundances by specific microbial communities. They comprise a unique pool of highly diverse, mostly uncultivable and host-specific microorganisms. Structure of the communities is strongly dependent on a-biotic and biotic factors. Associated bacteria have a function for nutrient supply and pathogen defence. Knowledge about diversity and function was used to isolate bacteria for biotechnological applications. The biodiversity of associated microbial communities can be characterized, by a combination of methods as given below:

a) Phylogenetic diversity
b) Spatial organization and specificity using clone libraries,
c) Fluorescent *in-situ* hybridization (FISH) and
d) SSCP fingerprinting.
e) Multivariate data analysis using *CANOCO* software or other platforms.

Metagenomic approaches have been used to find biosynthetic genes or compounds of interest for biocontrol and pesticides in agricultural systems (Table 17.1). Morgan and coworkers (2001) collected genetic material from three strains of *Xenorhabdus nematophilus*, a pathogen of cabbage white butterflies (*Pieris brassicae*) and isolated the biosynthetic genes and proteins associated with the insectidal activity. A fosmid library from the genetic material of *Penicillium coprobium* PF1169 was screened using colony PCR on small pools of clones to find the biosynthetic genes of an insecticide, pyripyroprene. With the knowledge of other pyripyropenes, primers were designed to target clones and then introduce the positive clone's vector into a model fungus, *Aspergillus oryzae*, for mechanistic studies (Hu *et al.* 2011). A third study focused on a bacterial pathogen, Serratia entomophila strain Mor4.1, of several soil pests from the *Phyllophaga* and *Anomala* genera. DNA was isolated, used to generate a fosmid library, and individual clones were injected into the insect larvae as a screening technique. Proteins in the cell membrane were found to be toxic and could provide a starting structure for biocontrol design efforts (Rodrý´guez-Segura *et al.* 2012).

Soil metagenome is a source for searching novel cry' genes and other toxin encoding genes. This will create a new revolution in microbial derived pest management strategies.

17.2. METAGENOMICS IN DISEASE MANAGEMENT

Two different metagenomic approaches to soil are possible, i.e. either unselective or targeted metagenomics. Unselective soil metagenomics constitutes a gene 'fishing' expedition, because no a priori selection (e.g. via prior growth-based selection of the communities under study, or following PCR amplification of a given gene or DNA region of interest) takes place before the metagenome is obtained and analyzed. Given the typical distributions of microbial species in most soils, hit rates of target genes are affected and can be low if targets occur in non dominant species. In this context, direct shotgun sequencing of the soil DNA pool bypasses the cloning step and instead relies on random high-throughput sequencing of soil DNA,or a target fraction thereof. By contrast, in a targeted soil metagenomic approach, the isolated pool of DNA is deliberately 'biased' to enhance hit rates, for instance owing to

(i) Pretreatment of the microbial community

(ii) Pre-fractionation of the DNA on the basis of G+C%

(iii) Amplification of target regions by PCR before metagenomic analysis.

Pretreatment consists of a pre-enrichment of target microbial groups by adding particular growth substrates, such as chitin, or a selection of particular cells by cell-sorting procedures.

17.3. METAGENOMICS AND THE STUDY OF PLANT GROWTH-PROMOTING RHIZOBACTERIA

The term PGPR (Kloepper and Schroth 1978) refers to those plant root ('rhizosphere')-associated bacteria that are capable of stimulating plant growth, e.g. by improving plant nutrition, by the production of plant growth regulators or by preventing the attack of pathogenic microorganisms. PGPR vary in their degree of intimacy with the plant, from intracellular, *i.e.* existing inside root cells, to extracellular, i.e. freeliving in the rhizosphere (Gray and Smith 2005). Some PGPR are commercially available as inoculants and have applicability for example in agriculture, forest regeneration and phytoremediation of soils (Lucy *et al.* 2004).There is a clear potential for metagenomics to contribute to the study of microbial communities of the rhizosphere, in particular PGPR. Possible contributions include (1) the discovery of novel plantgrowth promoting genes and gene products (2) the characterization of (not-yet-)culturable PGPRs. A high proportion of isolates show antifungal activities and a remarkable plant growth promotion by nitrogen fixation, phosphate solubilisation and production of phytohormons as well as an extraordinarily high biocontrol potential.

17.3.1. Novel plant growth-promoting genes and gene products

For many of the traits or mechanisms known to be plant growth-promoting, *in vitro* activity assays are applied to exploit gain-of function from a metagenomic library of rhizosphere DNA. For example, antibiotic activity towards (plant-pathogenic) bacteria or fungi can be assessed by testing whole-cell library clones or their extracts in a medium- or high-throughput manner for performance in confrontation assays. Many of such assays have been described using a variety of indicator strains, including some of the most important soil borne pathogens, e.g. the bacteria *Erwinia* (Emmert *et al.* 2004) and *Xanthomonas* (Rangarajan *et al.* 2003), the fungi *Fusarium* (Kim *et al.* 2006), and Rhizoctonia (Kim *et al.*2006; Rangarajan *et al.* 2003), and the fungus-like oomycetes, Phytophthora (Kim *et al.* 2006) and Pythium (Kim *et al.* 2006; Rajendran *et al.* 1998).

Production of the plant hormone indole 3-acetic acid (IAA) by metagenomic library clones can be measured using high-pressure liquid chromatography or colorimetric assays while cytokinins and their metabolites are detectable in supernatants by e.g. immunoaffinity chromatography. Genes for nitrogen fixation are retrievable with the use of nitrogen-free medium. Similarly, genes for the utilization of particular rhizosphere exudates could be recovered using an all-or-none complementation selection for growth on minimal medium containing these exudates as sole source of energy, carbon and/or nitrogen. In a similar approach, clones expressing 1-aminocyclopropane 1-carboxylate (ACC) deaminase, a plant growth promoting enzyme that lowers plant ethylene levels, could be selected by using ACC as sole source of nitrogen. The activities of lytic enzymes are most easily identified through clear zones around colonies on solid media, as has been documented e.g. for several biocontrol chitinases (Leveau *et al.* 2006). Assays based on halo formation are also available to identify PGPR-related phenotypes such as solubilization of mineral phosphate (Rodriguez *et al.* 2000) and siderophore production (Lee *et al.* 2003).

For the functional screening of library clones for PGPR functions, the use of alternative hosts seems very promising and rational. There is an abundant availability of phylogenetically diverse culturable PGPRs which could improve the probability of finding genes of interest, especially those that are not expressed in *E. coli* and whose full activity requires a specific PGPR background. The host role could also be played by several of the numerous defined mutants of PGPR that carry a knockout in one or several genes contributing to a particular PGPR phenotype. Such mutants are useful to screen libraries for heterologous genes and gene functions by a functional complementation approach. e.g. single-gene complementation of a mutant of *Burkholderia* sp. strain PsJN in quinolinate phosphoribosyltransferase (QAPRTase) activity (Wang *et al.* 2006), a mutant of *Pseudomonas putida* WCS358 unable to produce the antibiotic pseudobactin 358 (Devescovi *et al.* 2001), and a mutant of *Pseudomonas chlororaphis* PCL1391 impaired in the production of the antifungal secondary metabolite phenazine-1-carboxamide (PCN; Girard *et al.* 2006).

Mutants that lack one or more genes in a multi-gene pathway can be used for the production of antibiotics by enzymes such as polyketide synthases and non-ribosomal peptide synthases. The modularity underlying such proteins allows for a strategy of combinatorial complementation possibly leading to the discovery of antimicrobials with new structures and new target specificities. Activity screenings have the potential to retrieve novel genes with PGPR activity from the metagenomic pool. In contrast, the metagenomic harvest from sequence-based approaches such as PCR and/or Southern hybridizations will inevitably uncover only genes that match the specificity of the primers and/orprobes that were used to find them. Nevertheless, screening rhizosphere DNA for PGPR-related genes by PCR has several advantages in a metagenomic setting. Most importantly, sequencing the flanking regions of such genes on large-insert fosmid or BAC clones could provide insight into the identity of their owner, on the genetic context of these PGPR genes and possibly on the mechanisms of their regulation.

Highly complementary to activity- and sequence based screenings, a third approach to finding novel plant growth-promoting genes and gene functions is through comparative metagenomics. There is ample evidence that the microbial diversity measured by phylogenetic markers such as ribosomal RNA genes can differ dramatically between bulk and rhizosphere soil. Environmental gene tagging (EGT fingerprinting) by shotgun sequencing or suppressive subtractive hybridization of bulk and rhizosphere soil compartments could reveal differences in the type of gene adaptations that each compartment selects for. It is expected that genes with PGPR-like functions would be enriched in the rhizosphere library. Similarly, comparison of the genomic diversity of disease-suppressive and non-suppressive soils could expose genetic factors that contribute to or are predictive of the suppressiveness towards e.g. pathogenic microorganisms or nematodes.

17.3.2. Characterization of (not-yet-) culturable PGPRs

The knowledge on the abundance and activity of not-yet culturable PGPR is limited. However, there are several examples of their existence and contribution to plant health, e.g. *Pasteuria penetrans*, a not-yet-culturable bacterium is parasitic to plant-pathogenic nematodes, nitrogen fixing activity by viable-but-not-culturable grass endophytes the *Azoarcus* and the obligate biotrophism of arbuscular mycorrhizal (AM) fungi.

Bacteria belonging to the Acidobacteria and Verrucomicrobia are the most abundant, difficult-to-culture representatives in many rhizospheres. But, it is not clear if and how their abundance is correlated to their contribution towards plant health. A phylogenetic anchoring approach, would allow a (partial) insight into the genomes of these bacteria beyond the limited dataset that currently exists and into their possibly beneficial effect on plant growth. An analysis of the rhizosphere by comparative metagenomics holds the promise to reveal several important questions regarding the unculturable fraction of the rhizosphere community.

a) What actually constitutes this fraction from a comparison of metagenomic DNA isolated directly from rhizosphere to DNA isolated from all the colonies forming on solid media after plating from that same rhizosphere (i.e. the culturable fraction). With largescale DNA sequencing of both libraries, a start could be made to contrast the genetic diversity of the two populations.
b) Furthermore, by comparison of the functions enriched for in a library from rhizosphere soil versus bulk soil, the degree of the selection in each of the compartments for particular microbial activities, specifically those with PGPR relevance, can be estimated.

Shotgun sequencing in combination with a prior enrichment strategy towards an originally complex rhizosphere population allows the metagenomic analysis of rhizobacteria with a particular function of interest. Using this approach (RC-I) Archaea with origin in the rice rhizosphere was isolated from a methanogenic enrichment culture using rice paddy soil as an inoculum.

17.4. PHYTOPATHOGEN-SUPPRESSIVE SOIL

Phytopathogen suppressive soil does not contain a single activity of interest, rather it comprises a range that suppresses the growth and activity of plant pathogens including xenobiotic degradation,antibiosis, or antibiotic resistance.

17.4.1. METACONTROL

METACONTROL is a large project funded by the European Union was a collaborative effort of seven European labs including the small enterprise LibraGen (Toulouse, France), with the main aim of exploring the biotechnological potential of phytopathogen-suppressive soils. The underlying assumption was that the microbiota of these suppressive soils would serve as rich reservoirs of anti-phytopathogen loci, such as those involved in the production of antibiotics of the polyketide class and chitinase biosynthesis. In addition to one control soil, four different soils known to suppress pathogens were the source of metagenomic DNA for clone libraries. Clones were screened against known phytopathogens and observed for antibiosis activity resulting in, 0.05% positive clones across all libraries. Additional screening was done using primers specific for polyketide synthetase genes and positive hits increased to 0.22% (van Elsas *et al.* 2008b).

The project partners developed a range of methodologies that facilitated the exploration of the suppressive soil libraries. The technology used by the META CONTROL project partners is depicted in the figure 17.1:

17.5. EXPLORING DIVERSITY OF PLANT DNA VIRUSES

Viral metagenomics has been used to characterize viral communities present in the individual plants. The advantage of metagenomics for viral identification is that it allows for characterization of the complete viral community, including viruses with circular or linear genomes, and viruses that are too divergent to be detected by PCR assays based on known viral sequences. Purification of viral particles before sequencing ensures that the vast majority of the metagenomic sequences originate from viruses, in contrast to direct deep sequencing.

17.5.1. Vector-enabled metagenomics

VEM is a novel approach that characterizes the active and abundant viruses that produce disease symptoms in crops, as well as the less abundant viruses infecting adjacent native vegetation. The vector-enabled metagenomics (VEM) approach has been employed to study the highly polyphagous and mobile nature of the whitefly vector, combined with the capability of metagenomics to discover novel viruses without relying on sequence similarity to known viruses. VEM involves sampling of insect vectors from plants, followed by purification of viral particles and metagenomic sequencing. The VEM approach exploits the natural ability of highly mobile adult whiteflies to integrate viruses from many plants over time and space, and leverages the capability of metagenomics for discovering novel viruses.

17.5.1.1. *VEM for viral disease management*

PCR assays designed from the metagenomic sequences enabled the complete sequencing of four novel begomovirus genome components, as well as the first discovery of plant virus satellites in North America. One of the novel begomoviruses was subsequently identified in symptomatic *Chenopodium ambrosiodes* from the same field site, validating VEM as an effective method for proactive monitoring of plant viruses without a priori knowledge of the pathogens. VEM can be used to describe the circulating viral community in a given region, which will enhance the understanding of plant viral diversity, and facilitate emerging plant virus surveillance and management of viral diseases.

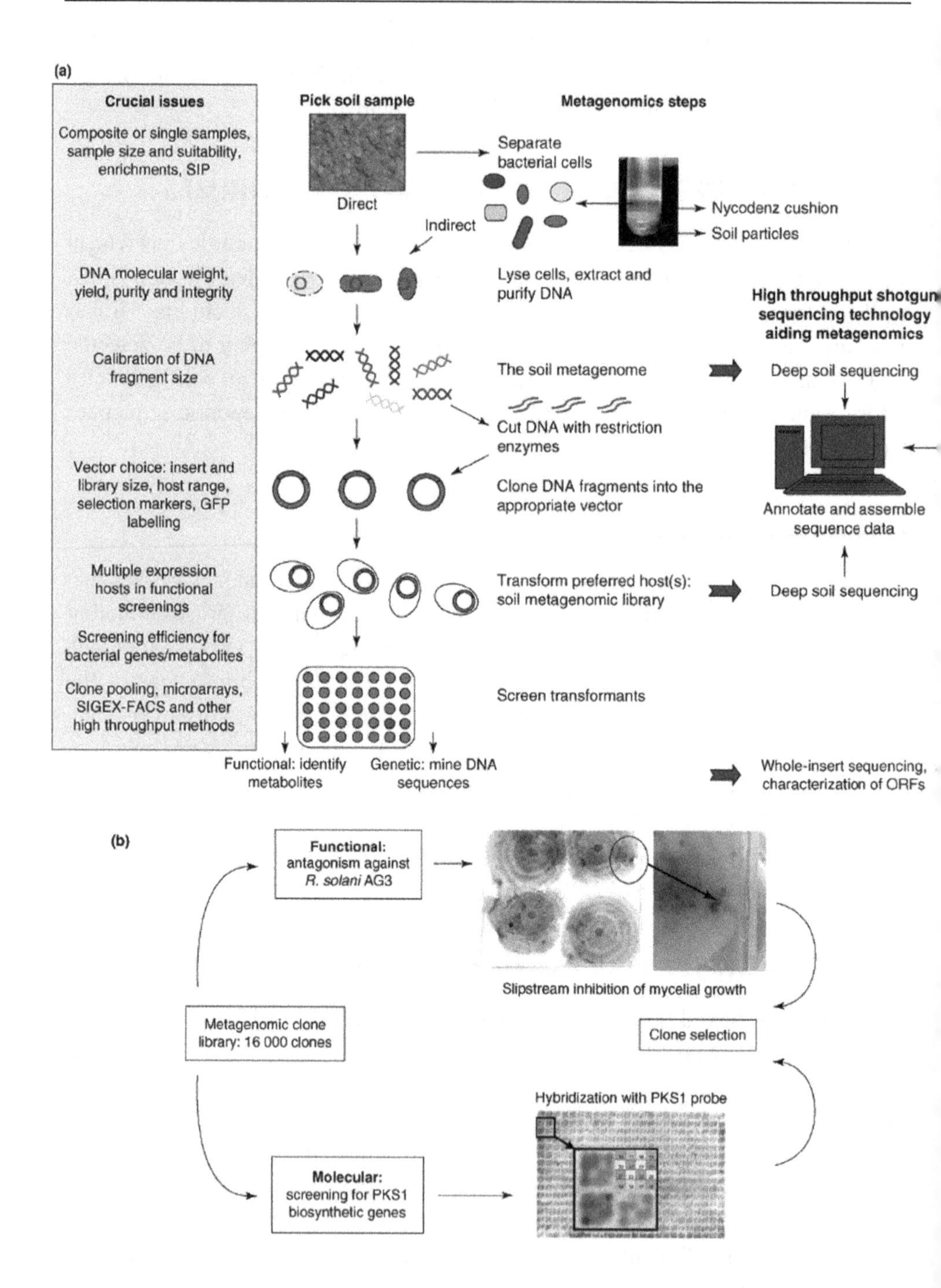

Fig. 17.1. Metagenomic exploration of Phytopathogen-suppressive soil

17.6. METAGENOMICS IN PLANT DISEASE DIAGNOSIS

Metagenomic studies have been developed to examine plant pathogens with the goal of finding the genes involved with pathogenicity. The earliest studies were on a virulence and xylanase deficient mutant of *Xanthomonas oryzae* pv. *oryzae* and wild-type *Xanthomonas campestris pv. vesicatoria*. While the study on *X. oryzae* pv. *oryzae* focused on virulence genes), the other study focused on finding the genes conveying hypersensitivity responses in tomato (*Lycopersicon esculentum L.)*. Virulence genes in pathogens such as *Phytophthora infestans*, *X. oryzae pv. oryzae* and *Ustilago hordei* were identified. The genes encoding secretion proteins important to toxin release were also unlocked. More expansive studies on compounds facilitating plant–microbe interactions have yet to be conducted.

A novel, unbiased approach to plant viral disease diagnosis has been developed which requires no a priori knowledge of the host or pathogen. Next-generation sequencing coupled with metagenomic analysis produced large quantities of cDNA sequence in a model system of tomato infected with Pepino mosaic virus. For instance, *Gomphrena globosa* infected with an unknown pathogen originally isolated from the flowering plant Liatris spicata was found to contain a new cucumovirus, named 'Gayfeather mild mottle virus'. This method expedites the entire process of novel virus discovery, identification, viral genome sequencing and, subsequently, the development of more routine assays for new viral pathogens.

The extraction of RNA from the infected plant, the production of cDNA with a random priming method and, finally, sequencing will produce sequences from a large range of potential pathogens. RNA viruses, viroids and the RNA stages of actively replicating DNA viruses can be directly sequenced. This approach should also produce sequences of mRNA and rRNA from any phytoplasma, bacteria or fungi present in the sample. By sequencing cDNA and not genomic DNA, only active host genes and ribosomes will be sequenced, avoiding the large amounts of untranscribed genomic DNA found in higher plants, and also avoiding integrated genomes of some plant viruses, such as badnaviruses.

To date, there have been three published accounts of the combination of next-generation sequencing and metagenomics in disease diagnostics. Cox-Foster *et al.* (2007) sequenced cDNA from a series of beehives with symptoms of colony collapse disorder and identified the presence of Israeli acute paralysis virus.

17.7. METAGENOMICS IN WEED MANAGEMENT

The massive diversity of bacterial and fungal genomes in soil provides a potentially vast pool of genes that code for the production of herbicidal compounds and herbicide resistance. As a consequence, two approaches to weed management can be developed using metagenomic techniques:

(1) Isolation of novel herbicides produced by vector–hosts expressing the biosynthesis genes

(2) Identification of herbicide resistance genes in vector–hosts exposed to high levels of an herbicide.

17.7.1. Developing Functional Screens for Weed Management

Countless screening assays specific to weed management can be developed using metagenomic clone libraries. Traditional assays for determining phytotoxic activity can be adopted for screens with vector–host systems. Additionally functional screening methods for discovering herbicide resistance genes are analogous to techniques used in the search for antibiotic resistance genes from soil.

17.7.1.1. *Isolating Novel Herbicides*

Although the process of developing metagenomic clone libraries remains a time intensive procedure, screening the libraries can be simpler and approachable. In order to screen individual clones, the first step to any functional screen is to dilute and plate the clones onto the relevant selective medium (Figure 17.2).

Invention of novel plating and separation techniques in this part of the assay should work on optimizing a primary screen for clones with secreted compounds or those of interest to the researcher. Separated clones are grown in large, liquid volumes and can be screened as diluted cell suspensions, supernatants, or crude extracts. In all situations, background effects of the microbial cell or metabolites must be monitored via controls. Plant toxicity protocols have been standardized by both American and International organizations, and are easily adapted to functional screens. Established protocols include standardization of experimental conditions, recommendations for plant species to use, and many possible response measurements (Figure 17.2). Data are presented as a percent effect difference from a control plant, linear or nonlinear regression analyses, or tabulated seedling/root lengths and dry mass. Percent of germination at endpoint sampling also is another relevant measurement in addition to determining the lowest concentration with herbicidal effect and the highest concentration with no herbicidal effect.

From these observations, concentrations of unknown compounds able to decrease growth of a plant by 50%, EC_{50}, can be determined. Recent research commonly uses this information to test the effect of compounds on model plant species such as rice, lettuce, and cucumber in addition to plants of interest to the researcher.

Greenhouse studies are impractical as an initial screening assay; the frequency of candidate clones has been 1% in other screens for natural products. To increase the likelihood of finding herbicidal compounds, a high-throughput screen will provide the most success. Germination or early seedling assays can be adapted to a 96-well plate which allows for rapid dilution and application of clone cells, supernatants, or extracts in a replicable manner.

17.7.1.2. *Bioassay of Duckweed (Lemna minor L. or Lemna gibba L.) and algae*

Duckweed is a small, aquatic plant with small fronds, leaf-like structures, able to be grown in a 96-well format. Frond number, size, and color are assayed using a dissecting microscope or digital image analyzer. Another bioassay adapted to the 96-well format is based on a cell suspension of algae. A variety of algal species have been used including *Scenedesmus subspicatus*, *Pseudokirchneriella subcapitata*, *Chlamydomonas reinhardii*, and *Chlorella pyrenoidosa*. As higher amounts of a phytotoxic compound are added to an algal culture, more of the cells die, resulting in lower dry biomass. Using algal bioassays, traditional toxicity measurements such as EC_{50} and regression equations have been calculated for a variety of known herbicides.

17.7.1.3. *Identifying herbicide resistance genes*

The use of GM crops has become more prevalent. Herbicide resistance genes from soil microorganisms provide a valuable source for engineering new GM crops resistant to herbicides. Similar to the search for antibiotic resistance genes and mechanisms, *E. coli* clones can be grown on a medium containing the phytotoxic or herbicidal compound. Colony growth on the medium could indicate the presence of a gene or gene cluster either resistant to the herbicide or able to metabolize the herbicidal compound. Another screen can be developed using herbicidal compounds modified with colorimetric additions to indicate any excessive changes in the host's metabolism. Eg) glyphosate resistance and glyphosate degrading abilities using microorganisms.

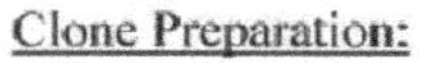

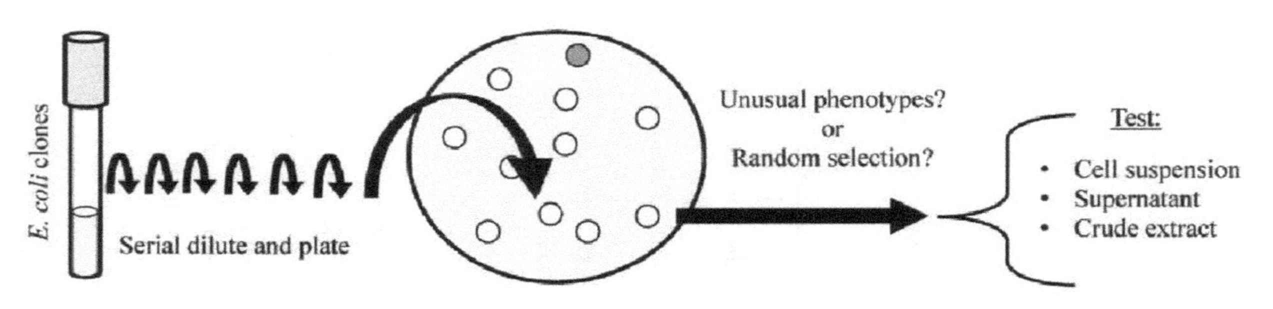

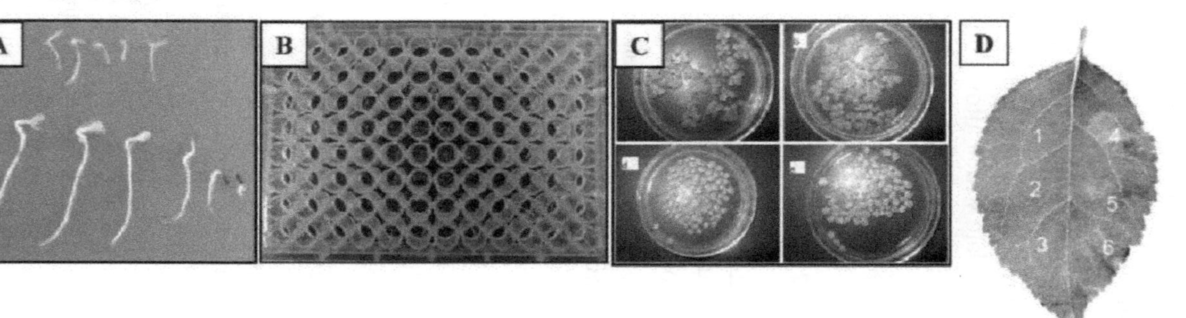

Fig. 17.2. Schematic showing preparation of clone libraries and the screening assays for herbicide detection

If the environmental gene is from a eukaryotic organism,the gene will not likely be expressed in E. coli. Therefore, choosing a vector to allow transfer between host organisms or even to screen the library in a eukaryotic host, e.g. Saccharomyces cerevisiae, will increase the chance of expression.Hosts that are filamentous fungi, in the genera Aspergillus or Trichoderma, are also available. Once herbicidal resistance genes are found, transfer of the genes into Agrobacterium tumefaciens would allow for introduction into model plants and extended physiology studies on the mechanisms involved with resistance. Prior knowledge of the plant physiological responses to a given herbicide could assist with functional screening designs for discovering mechanisms of resistance.

17.8. METAGENOMICS IN CLIMATE CHANGE

Metagenomics approaches unravel the potential microbes to oxidize methane in flooded ecosystem which inturn influences global warming. Mining beneficial alleles/genes for stress tolerance viz., drought, temperatue, cold and salinity is another area of expedition using metagenomics coupled with metatransciptomics and metaproteomics. Introgression of these genes into crop plants will open new vistas for 'evergreen revolution' to happen.

17.8.1. Structure of belowground microbial communities at elevated CO_2

Understanding the responses of biological communities to elevated CO_2 (eCO_2) is a central issue in ecology, but little is known about the influence of eCO_2 on the structure and functioning (and consequent feedbacks to plant productivity) of the belowground microbial community. Metagenomic technologies showed that eCO_2 dramatically altered the structure and functional potential of soil microbial communities. Total microbial and bacterial biomass is significantly increased at eCO_2, but fungal biomass is unaffected. While the abundance of genes involved in decomposing recalcitrant C remains unchanged, those involved in labile C degradation and C and N fixation significantly increases under eCO_2.

17.9. SPECIES ABUNDANCE DISTRIBUTIONS AND RICHNESS ESTIMATIONS IN FUNGAL DIVERSITY

The diversity and community ecology studies strongly depend on sampling depth. Completely surveyed communities follow log-normal distribution, whereas power law functions best describe incompletely censused communities. It is arguable whether the statistics behind those theories can be applied to voluminous next generation sequencing data in microbiology by treating individual DNA sequences as counts of molecular taxonomic units (MOTUs). The suitability of species abundance models in three groups of plant-associated fungal communities phyllosphere, ectomycorrhizal and arbuscular mycorrhizal fungi were reported. The impact of differential treatment of molecular singletons on observed and estimated species richness and species abundance distribution models were also studied. The arbuscular mycorrhizal community of 48 MOTUs followed log-normal distribution. The ectomycorrhizal (153 MOTUs) and phyllosphere (327 MOTUs) communities significantly differed from log-normal distribution. The analysis of abundant (core) and rare (satellite) MOTUs clearly identified two species abundance distributions in the phyllosphere data – a log-normal model for the core group and a log-series model for the satellite group. The prominent log-series distribution of satellite phyllosphere fungi highlighted the ecological significance of an infrequent fungal component in the phyllosphere community.

17.10. METAGENOMICS APPLIED TO ARBUSCULAR MYCORRHIZAL FUNGAL COMMUNITIES

Metagenomics studies have recently offered new approaches that shed light on microbial communities in a variety of environments. In this context, DNA pyrosequencing is being used more and more to investigate prokaryotic assemblages in soil environment. Fungi, which are crucial components of soil microbial communities, functioning as decomposers, pathogens and mycorrhizal symbionts, have instead been largely neglected. At the moment, only two studies have investigated AMF biodiversity using the pyrosequencing approach and SSU rDNA as the target gene. Although both studies targeted the same group of fungi, they focused on different habitats. Compared to similar studies, carried out using a cloning-sequencing approach or DNA barcoding, the main outcome that emerged from pyrosequencing analyses applied to AMF and to other fungal communities is the unexpected fungal biodiversity observed in the analyzed environments. Interestingly, the pyrosequencing approach applied to isolated spores of AMF has demonstrated that they are a niche for highly polymorphic endobacterial communities. The data confirm the powerfulness of the pyrosequencing approach, which represents a promising new tool to better understand the natural distribution of an essential group of soil microorganisms, such as fungi. The large number of reads that have been obtained increases the likelihood of capturing sequences from rare organisms, which would instead remain undetected with the cloning-sequencing approach.

17.11. SCOPE OF METAGENOMICS IN ENDOPHYTES

Endophytes are microorganisms that live within plant tissues without causing symptoms of disease. They are important components of plant microbiomes. Endophytes interact with, and overlap in function with, other coremicrobial groups that colonize plant tissues, e.g., mycorrhizal fungi, pathogens, epiphytes, and saprotrophs. Control of insects and fungal pathogens can be accomplished by different mechanisms: by antagonism, including competition, parasitism, or production of secondary metabolites; by induction of host defenses; or by stimulation of host growth and vigor. These mechanisms overlap, and a single endophyte may employ several of them. Most studies have proposed the induction of host defenses, specifically mechanisms of systemic acquired resistance (SAR). SAR is often mediated by production of salicylic acid, jasmonic acid, ethylene, and a variety of pathogenesis-related (PR) proteins. Endophytes may also produce secondary metabolites that directly inhibit insects or pathogens, or produce elicitors that stimulate the plant to produce secondary metabolites. Furthermore, a single endophyte may offer protection from both fungal pathogens and insects.For example, *Beauveria bassiana* inhibits both fungal pathogens and insects, mostly by production of secondary metabolites, and *Lecanicillium* spp. and *Trichoderma* spp. are both mycoparasites and insect parasites, although they also produce inhibitory metabolites. Endophytes also find its application in phytoremediation. Hidden from view and often overlooked, endophytes are emerging as their diversity, importance for plant growth and survival, and interactions with other organisms are revealed. The bottleneck in exploring endophytes is its unculturable nature. Metagenomics tool will decipher endophyes and other related microbiomes for its application in Agriculture.

17.12. NITROGENASE GENE DIVERSITY

Biological nitrogen fixation is an important source of fixed nitrogen for the biosphere. Microorganisms catalyse biological nitrogen fixation with the enzyme nitrogenase, which has been highly conserved through evolution. Cloning and sequencing of one of the nitrogenase structural genes, *nifH*, has provided a large, rapidly expanding database of sequences from diverse terrestrial and aquatic environmentrs. Comparison of *nifH* phylogenies to ribosomal RNA phylogenies from cultivated microorganisms shows, little conclusive evidence of lateral gene transfer. The phylogeny of nitrogenase includes branches that represent phylotypic groupings based on ribosomal RNA phylogeny and also paralogous clades including alternative, non-molybdenum, non-vanadium containing nitrogenases. Only a few archaeal nitrogenase sequences have been obtained from the environment. Extensive analysis

Table 17.1 Metagenomic studies relevant to the agricultural sciences. The targeted genes included in the list were isolated from large-insert clone libraries. "ns" refers to "not specified".

Origin	Vector	Average insertsize kb	Genes of interest	Reference
Xanthomonas oryzae pv. oryzae (virulence and xylanase deficient mutant)	Cosmid	ns	Genes for virulence and xylanase activity	Ray *et al.* 2000
Xanthomonas campestris pv. vesicatoria	Cosmid	ns	Genes conveying hypersensitive responses in Lycopersicon esculentum	Astua-Monge *et al.* 2000
Phytophthora infestans	BAC	ns	Genes conveying avirulence to oomycete plant pathogen	Whisson *et al.* 2001
Xanthomonas oryzae pv. pryzae	BAC	107	Variety of genes encoding avirulence, hypersensitivity, pathogeneity, and virulence	Ochiai *et al.* 2001
Xenorhabdus nematophilus PMF1296	Cosmid	ns	Insecticide (against Pieris brassicae larvae)	Morgan *et al.* 2001
Mycobacterium sp. strain ESD	Cosmid	ns	Endosulfan degrading protein	Sutherland *et al.* 2002
Pseudomonas syringae	Cosmid	ns	Secretion genes	van Dijk *et al.* 2002
Erwinia carotovora subsp. Atroseptica	BAC	ns	Secretion and pathogenicity genes	Bell *et al.* 2002
Uncultivable phytoplasma with pathogenic activity	Cosmid	19	Whole genome	Liefting and Kirkpatrick 2003
8 different strains of Ustilago hordei	BAC; Cosmid	ns	Genes encoding avirulence	Linning *et al.* 2004
Alternaria alternata	Cosmid	ns	Toxin synthesis genes	Ito *et al.* 2004

[Table Contd.

Contd. Table]

Origin	Vector	Average insertsize kb	Genes of interest	Reference
Phytophthora nicotianae	BAC	90	HSP genes	Shan and Hardham 2004
Tribolium castaneum stain GA-2	BAC	129	Chitin synthase	Arakane *et al.* 2004
Rhodococcus opacus TKN14	Cosmid	Ns	o-xylene oxygenase	Ns Maruyama *et al.* 2005
Activated sludge	Fosmid	33	Extradiodioxygenases	Suenaga *et al.* 2007
Penicillium coprobium PF1169	Fosmid	ns cluster	Pyripyropene A biosynthetic gene	Hu *et al.* 2011
Serratia entomophila Mor4.1 culture	Fosmid	40	Toxicity for Phyllophaga blanchardi larvae	Rodrý´guez-Segura *et al.* 2012

of the distribution of nifH phylotypes among habitats indicates that there are characteristic patterns of nitrogen fixing microorganisms in termite guts, sediment and soil environments, estuaries and salt marshes, and oligotrophic oceans. The distribution of nitrogen fixing microorganisms, although not entirely dictated by the nitrogen availability in the environment, is non random and can be predicted on the basis of habitat characteristics (Zehr *et al.,* 2003).

17.13. FUTURE THRUST

In summary, the tools of metagenomics offer many openings into a broadened view of the rhizosphere. Probably the biggest obstacle in the construction of a metagenomic library from rhizosphere soil DNA is the relative low availability of starting material. To 1 cm of root typically adheres only 20 mg of soil, so one needs (depending on the plant species under investigation) 50 to 500 cm of root material in order to apply a DNA extraction method that requires 1 to 10 g of soil. Several protocols have also been developed for the isolation of metagenomic bacterial DNA from inside plant material.

It will also benefit our ability to improve existing PGPR, by adding to the pool of exploitable PGPR genes and utilization of this pool to develop PGPRs with enhanced performance. The discovery of novel PGPR activities, either by functional screening or based on DNA sequence information, will add enormously to our understanding of the mechanistic variation that exists in PGPR phenotypes.

The use of metagenomics in parallel with established or novel molecular approaches to he study of PGPR, such as genome sequencing of new PGPR isolates will undoubtedly lead to the discovery of novel mechanisms of PGPR activity, new types of PGPR identity and a fresh look on the biology and practical application of PGPR.

Still, a number of technical challenges are pending. Fungal metagenome studies are limited because with current technologies plant DNA cannot be separated from fungal DNA; the high concentration of host plant DNA makes the much larger but much more dilute fungal metagenome difficult to sequence with adequate coverage.

We are also limited by our capacity to manage large datasets and to conduct automated classification, and by insufficient functional gene annotations for current genomes.

Endophytes are a mine for bioprospecting. By extrapolation from numbers of new endophytes and metabolites reported in the past twenty years, many interesting cases are yet to be discovered.

Future Issues

1. Toward a synthesis era: Can the research community agree on guidelines for isolation, extraction, amplification, annotation, and sequence database curation?
2. How can we best apply a systems biology approach to complex, simultaneous interactions of the different groups of organisms that affect plant health?
3. How can useful endophytes be applied for pest control or to confer other desired traits on their hosts?

Mini Quiz

1. Explain the significance of metagenomics in biocontrol?
2. What is the importance of *Pasteuria penetrans*?
3. Write about METACONTROL project?
4. What is the role of metagenomics in weed management?
5. Define species abundance and richness?

CHAPTER - 18

BIOINFORMATICS IN METAGENOMICS

Metagenomics has become an indispensable tool for studying the diversity and metabolic potential of environmental microbes that are not yet cultured. Continual progress in next-generation sequencing allows for generating increasingly large metagenomes and studying multiple metagenomes over time or space. Recently, a new type of holistic ecosystem study has emerged that seeks to combine metagenomics with biodiversity, meta-expression and contextual data. Such 'ecosystems biology' approaches bear the potential to not only advance our understanding of environmental microbes to a new level but also impose challenges due to increasing data complexities, in particular with respect to bioinformatic post-processing. This chapter addresses selected opportunities and challenges of modern metagenomics from a bioinformatics perspective.

18.1. BIODIVERSITY AND BIOINFORMATICS

The development of techniques for sequencing deoxyribonucleic acid (DNA) from environmental samples is a crucial factor for the discovery of diversity among prokaryotes. In particular, techniques to obtain 16S ribosomal ribonucleic acid (rRNA) sequences from the environment, such as the early reverse transcriptase-based approaches and the later polymerase chain reaction-based methods have been cornerstones toward current large-scale studies of microbial biodiversity. More than 3 million 16S rRNA sequences of Bacteria and Archaea in the release 111 of the SILVA database constitute an impressive hallmark of microbial versatility. It represents just a fraction of the diversity of soils where just a single ton is believed to potentially harbor millions of species. With respect to microbial diversity we so far have seen just the proverbial tip of the iceberg. The classical metagenome approach involves cloning of environmental DNA into vectors with the help of ultra-competent bioengineered host strains. The resulting clone libraries are subsequently screened either for dedicated marker genes (sequence-driven approach) or metabolic functions (function-driven approach). Nowadays, direct sequencing of environmental DNA (shotgun metagenomics) is commonly used to study the gene inventories of microbial communities. By combining the resulting metagenomic data with biodiversity data (e.g. from 16S rRNA gene amplicon sequencing in situ expression data (metatranscriptomics and metaproteomics) and environmental parameters, a new type of holistic ecosystem studies has become feasible (Figure 18.1). Similarly, metagenome data can be integrated with metabolome data. Such integrative 'ecosystems biology' studies introduce a plethora of challenges with respect to experimental design and bioinformatic downstream processing. These involve considerations about the habitat, sampling strategy, sequencing technology, assembly, gene prediction, taxonomic classification and binning, biodiversity estimation, function predictions and analyses, data integration and subsequent interpretation and data deposition.

18.2. HABITAT

The biodiversity composition (richness and evenness) of a habitat has a profound impact on the quality of a metagenome. For metagenome analyses involving assembly (to generate longer genome fragments with multiple genes), habitats with few microbial species or an uneven population with few dominating species are more promising targets than habitats with many species of even abundance. However, more important than the absolute number of species is their level of genomic coherence. Even seemingly ideal habitats with a stable composition of few dominant species, for example microbial mats or invertebrate bacterial symbioses can be difficult to assemble when evolutionary micro-niche adaptations have led to large pan-genomes and thus to a low level of population clonality. In contrast, seemingly unsuitable habitats that harbor a multitude of species with dynamically changing compositions can yield long assemblies, when the species that thrive and dominate are largely clonal. This effect is observed when a second round of sequencing and reassembly of an environmental sample breaks rather than elongates assemblies from the previous round. The reason is that in habitats with little clonality, more sequencing covers more genomic heterogeneity. This increases incongruities in putative assemblies, which causes assemblers to generate smaller but congruent assemblies rather than long assemblies with high levels of positional variability. These common issues in metagenomics might be overcome by switching either to a longer read sequencing technology or by more sequencing

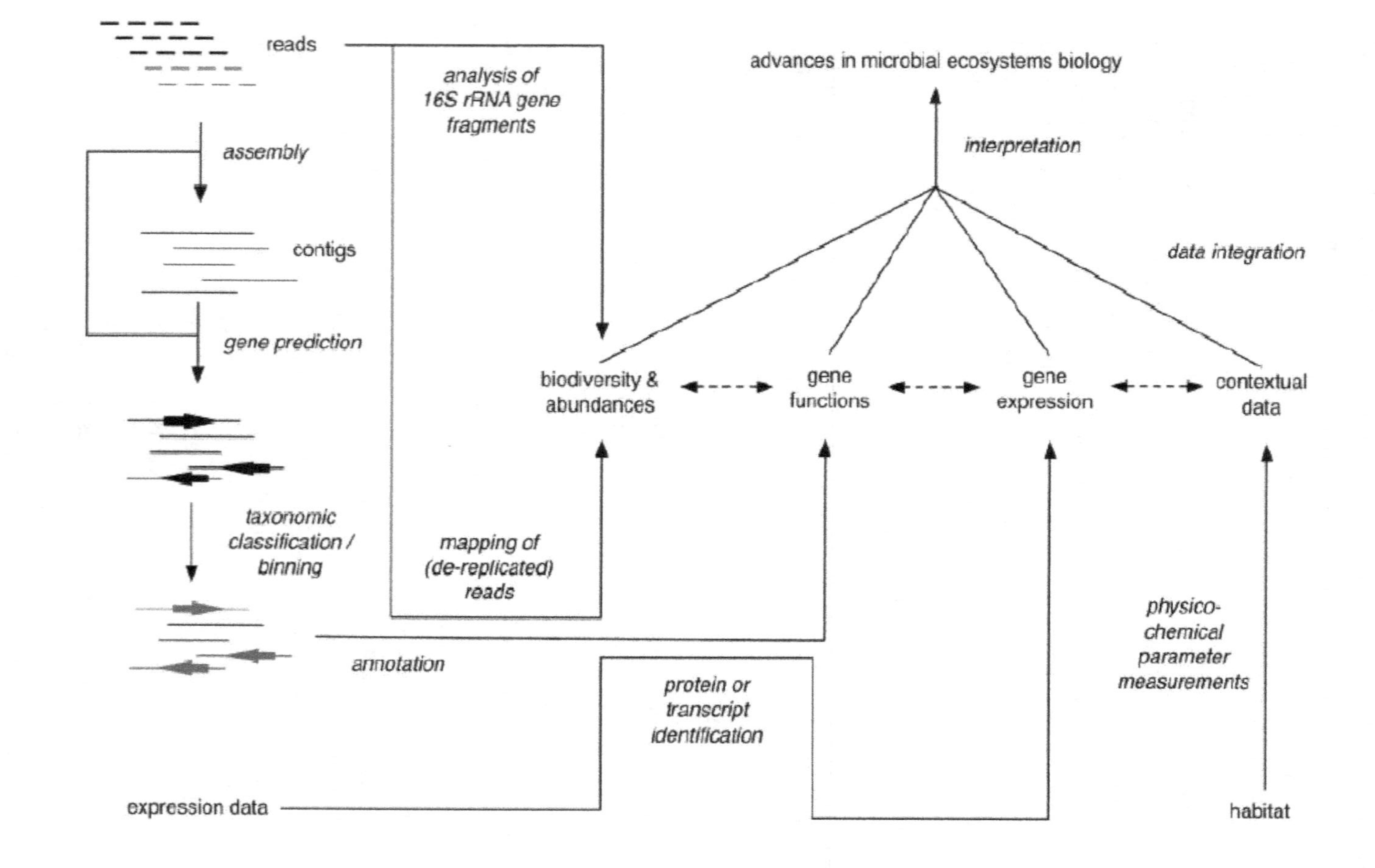

Fig.18.1. Scheme of the major stages of an integrative metagenomic ecosystems study on microbial ecology

to increase coverage. Preceding biodiversity analysis can help to properly assess the required amount of sequencing, e.g. in form of full-length 16S rRNA clone libraries analysis (for fine-scale resolution) in conjunction with 16S rRNA gene tag analysis (for abundance estimations).

18.3. SAMPLING

When the study target is a specific uncultivable microbial species of low abundance, it is worthwhile to try to enrich the species after sampling. Sometimes favorable culture conditions can be found that result into co-cultures with substantially enriched target species or the species can be physically enriched, for example by methods such as fluorescence activated cell sorting or by density gradient centrifugation. Subsequent multiple-displacement amplification and single cell sequencing might be a viable solution to obtain a draft genome. However, when the target is a habitat's overall function, a representative sample must be studied, and data from single cells, enrichments or isolates—though valuable—are complementary.

Sampling of microbes from environments usually involves a size selection (e.g. fractionating filtration) to minimize contaminations by viruses or eukaryotes. Such reduction of a sample's complexity introduces a bias in the community composition. It can be reasonable to reduce the complexity of an environmental sample by enrichment or size selection, in particular when multiple different enrichments or samples with different size fractions are taken and their results are combined, but such effects need to be taken into account in experiment design and interpretation of the final data.

18.4. REPLICATION

It is good scientific practice to analyze true replicates of a sample and to assess whether observed differences within one sample are statistically meaningful. However, this is rarely done in microbial ecology. One reason is that in many habitats it is almost impossible to take true replicates. Eg. Water samples that were taken within few minutes might differ because the sampled water was moving and not perfectly homogeneous. Hence, comparing such alleged replicates reveal little information on methodological reproducibility. Thus it is understandable that environmental biologists prefer to analyze more samples rather than to invest in replication, in particular in expensive large-scale projects. This, however, does not release scientists from assessing the reproducibility of their methods.

Pseudo-replication: The reproducibility can be addressed by pseudo-replication such as sub-sample analysis and comparison of samples within time series and by independent assessments of methodological reproducibility using representative test data sets.

18.5. SEQUENCING

Next-generation sequencing (NGS) has been nothing less than a paradigm shift for metagenomics. Not long ago, the classical clone-based metagenome approach in combination with Sanger sequencing usually allowed for obtaining only few selected inserts, as sequencing was the limiting factor. NGS has obliterated the cloning step and its inherent problems and enabled to sequence environmental DNA directly. Initially, 454/Roche pyrosequencing was most widely used, because it generated substantially longer reads than competing platforms. Meanwhile, in particular large-scale metagenome projects make increasing use of the Illumina and, to a lesser extent, SOLiD platforms. Although the latter two still provide shorter reads than pyrosequencing, they offer a much higher throughput and hence coverage for the same price. The read length of 2 x150 bp provided by the current Illumina GA IIx line of instruments basically matches that of the first generation 454 Life Sciences GS20 instrument and high coverage in conjunction with mate-pair libraries facilitate assembly and can compensate for the lack of read length. In addition, protocols have been proposed that allow for obtaining longer 'composite reads' from short read platforms.

It remains to be seen what impact newer sequencing platforms will have on the metagenomic field. The lately announced Ion Torrent Proton fits in between the 454/Roche and Illumina platforms in terms of read length and throughput at a seemingly competitive price. At the same time, single-molecule detection methods such as Pacific Bioscience's PacBio RS and the recently announced Oxford Nanopore Technologies (ONT) GridION and MinION systems offer much longer read lengths, albeit at the expense of higher sequencing errors (PacBio RS: 10–15% and ONT GridION: 4%). However, in contrast to the inherent systematic errors of other platforms, these errors are mostly random, and once these platforms improve, they could be reduced in an almost linear fashion by increased multifold sequencing of the exact same DNA molecule and increased coverage.

PacBio and ONT sequencing: The specific value of PacBio and ONT sequencing is that they provide read lengths that are long enough to span multiple prokaryote genes and thus are able to provide reliable genetic contexts. Currently, a combination of long- and short-read technologies constitutes a particularly promising approach in future metagenomics that bears the potential to significantly advance the field.

Table 18.1. Genome sequencing technologies

Platform	Method	Advantages	Disadvantages	System cost (cost per run)
First generation	Sanger sequencing			
Sanger	Strands of fragmented DNA are resolved on gel and distributed in order of length, with end base labeled	High accuracy Validate findings of NGS	High cost Low throughput (time consuming)	US$2,000,000
Second generation	Cyclic array-based sequencing; strands of fragmented DNA are amplified; bases are added sequentially using DNA ploymerase, excess reagent is washed out, imaging identities base incorporated; and process repeats	Higher throughput More economical	Short red length Complex sample preparation Need for amplication Long time to results Significant data shortage and interpretation requirements	
Hiseq* (Illumina, CA, USA)	Fluorescent-labeled nudeotides added simultaneously	<1 μg DNA needed	75 (35-100) bp read lengths More false positives	~US$400,000 (US$8950)
Miseq (Illumina)		Clinical applications	Unable for WES, WGS, ChIP-Seq and RNA-seq 10 h per run	
Roche Applied Science 454* genome sequencer (Roche, Basel, Switzerland)	Pytrophosphate releaset at time of base incorporation	1-5 μg DNA needed	400 bp read lengths	US$525,000 (US$8439)
Roche 454* GS Junior (Roche, Basel, Switzerland)		Clinical applications	Unable for WES, WGS, ChIP-Seq and RNA-seq 8 h per run	

[Table Contd.

Contd. Table]

Platform	Method	Advantages	Disadvantages	System cost (cost per run)
Applied Biosystems SOLID* 4 (Life Technologies, CA, USA)	Driven by DNA ligase instead of DNA polymerase	2-20 μg DNA needed	35-50 bp read lengths	US$50,000 (US$17,447)
Complete Genomics*		Clinical applications	35 bp read lengths Lacks sensitivity	Not commercially available
Ion Torrent PGM (Life Technologies)	Nonoptical DNA sequencing; massively parallel semiconductor senses ions produced as nucleotides are incorporated by DNA polymerase-based synthesis	Less than 200 bases needed High accuracy Short run time (fast) Cheaper	Unable for WES, WGS, ChIP-Seq and RNA-seq	US$50,000 (<US$500)
Ion Proton (Life Technologies)		2 h per run	Capable for WES, WGS Higher throughput than Ion Torrent	
Third generation	Novel technologies	No PCR amplification Less starting material Less error prone		
PacBio RS (Pracific Bio Science, CA, USA)	Single-molecule real-time sequencing, imaging of dye-labeled nucleotides as they are incorporated during DNA synthesis by single DNA polymerase molecule	800-1000 bp read lengths		US$695,000 (~US$1000)

[Table Contd.

Contd. Table]

Platform	Method	Advantages	Disadvantages	System cost (cost per run)
Heliscope sequencer (Helicos Bio Sciences)	Single-molecule real-time sequencing; imaging of dye-labeled nucleotides as they are incorporated during DNA synthesis by single DNA polymerase molecule	<2 μg DNA Direct RNA-sequencing application	35 bp read lengths	US$750,000 (~US$5,000)
Fourth generation				
Oxford Nanopore	Single molecule sequencing incorporating nanopore technology	Whole-genome scane 15 min Very low cost		Not commercially available

ChIP-Seq: Chromatin immunoprecipitation-sequencing; NGS: Next generation sequencing, RNA-seq: RNA-sequencing; WES: Whole-exome sequencing; WGS: Whole-genome sequencing.

18.6. ASSEMBLY

An assembly yields larger genomic fragments that allow for the study of gene arrangements. Valuable functional knowledge can be deduced from gene neighborhoods, e.g. when a gene of unknown function always appears together with a gene whose function is well known. Large-scale systematic investigations of such gene syntenies across metagenomes have the potential to uncover as yet unknown functional couplings. Assembly of sequences from metagenomic libraries can result in good draft or even complete genomes when the target species shows little intra-species variation, but this usually requires a substantial amount of sequencing. For example, massive sequencing allowed to obtain a good draft genome of *'Candidatus Cloacamonas acidaminovorans'* from a wastewater anaerobic digester. From a cow rumen metagenome 15 draft genomes were obtained by direct assembly.

Although assembly does yield longer sequences, it also bears the risk of creating chimeric contigs, in particular in habitats with closely related species or highly conserved sequences that occur across species (for example as a result from high transposase, phage and lateral gene transfer activities). Furthermore, assembly distorts abundance information, as overlapping sequences from an abundant species will be identified as belonging to the same genome and consequently joined. This leads to a relative underrepresentation of sequences of abundant species. Hence, gene frequencies are better compared based on read representation rather than on the basis of assemblages. An alternative is to back-trace all reads that constitute a given contig (or gene), either by direct mapping of the reads on the assemblies or by extracting the respective information.

18.6.1. Challenges in assembly and assemblers

a) **Chimeric assemblies:** Assemblers yield similar results when the coverage is high, but at low coverage, the assembler and its settings can have a notable effect. Furthermore, metagenome assemblers need to be more fault tolerant than genome assemblers to account for strain-level genomic heterogeneity, which on the other hand elevates the risk for *chimeric assemblies*. The metagenome assemblers that try to address these problems are Genovo, Meta-IDBA, MetaVelvet and MAP. The first three of these are intended for short-read data, whereas MAP also handles longer reads as they are produced by current 454 FLXþ pyrosequencers. All four assemblers are claimed to yield longer assemblies and more representative taxonomic representations than conventional assemblers.

b) **Memory:** A dedicated stand-alone metagenome scaffolder that can be used to post-process the unitig graphs of other assemblers is BAMBUS2. One particular problem is that the increasing throughput of NGS platforms imposes challenges on assembly, in particular with respect to *memory requirements*. A current Illumina HiSeq2000 sequencer can generate 600 Gb in a single run, and higher throughput technologies are almost given in the nearer future. As a result, metagenomics is currently experiencing a split between smaller, more targeted projects with assemblies and large-scale projects without assemblies. The trend in metagenomics for tremendous data scales has been anticipated even before second- and third-generation NGS platforms became available and has been termed 'megagenomics' . Example: the Human Microbiome and Earth Microbiome Projects.

18.7. GENE PREDICTION

Many conventional gene finders require longer stretches of sequence to discriminate coding from non-coding sequences and training sequences from a single species that is subsequently used to build a species-specific gene prediction model. This is unsuitable for metagenomes that are constituted as a mixture of sequences from different organisms and often comprise only a limited number of long contigs but mainly short assemblies and unassembled reads. The partial genes must be predicted missing proper gene starts, stops or even both. In addition, metagenomes (in particular at low coverage) are often riddled with frame shifts. This makes gene prediction for metagenomes a non-trivial task.

Gene prediction programs have been developed for metagenomes, such as MetaGene, MetaGeneAnnotator, Orphelia and FragGeneScan. All these programs have been built for short reads, but they follow different approaches (such as machine learning techniques and Markov models), differ in the precision of ribosomal binding site and thus correct start prediction and in their tolerance for sequencing errors. Combining multiple gene finders, screening intergenic regions for overlooked genes and using dedicated frameshift detectors [54, 55] are common strategies to overcome at least some of these limitations.

18.8. TAXONOMIC CLASSIFICATION AND BINNING

One of the key problems of current metagenomics is to assign the obtained sequences and their gene functions to dedicated taxa in the habitat. Phylogenetic marker genes are sparse and thus allow only taxonomic assignment of a minor portion of sequences. Hence, other approaches are needed that can partition metagenomes into taxonomically distinct bins (taxobins) that provide taxon-specific gene inventories with ecologically indicative functions. A number of such approaches have been developed that can be categorized into classification and binning approaches.

Classification approaches assign taxonomies based on similarities between metagenomic sequences and sequences of known taxonomy. Binning approaches work intrinsically (i.e. without reference sequences) and cluster sequences based on compositional characteristics. In general, one can discriminate methods that operate on the level of protein sequences (gene-based classification), on the level of intrinsic DNA characteristics (signature-based binning/classification) and those that map DNA reads to reference sequences (mapping-based classification).

18.8.1. Gene-based classification

Gene-based classification requires all the metagenomic sequences' potentially full and partial protein-coding regions to be translated into their corresponding protein sequences. There are two main approaches.

i) The first is to use conventional basic local alignment search tool (BLAST) searches against protein databases such as the non-redundant NCBI database or UniProt and to derive taxonomic information from the resulting hits. This can be done either by constructing a multiple sequence alignment from the best matching hits with subsequent phylogenetic reconstruction, as implemented in Phylogena or it can be done directly based on the BLAST results.

   BLAST is a heuristic for fast sequence database searches and not a phylogenetic algorithm per se, the top BLAST hit does not necessarily agree with the taxonomic affiliation of the gene in question. However, it has been shown that post-processing a larger number of BLAST hits can reveal useful taxonomic assignments, for example by using consensus information as implemented in the lowest common ancestor algorithm of MEGAN or the Darkhorse and Kirsten algorithms.

ii) A second possibility is to infer taxonomic information from HMMer searches against Pfam models as implemented in CARMA or Treephyler The principle of CARMA is to align a sequence hitting a Pfam model with the model's curated seed alignment, construct a neighbor-joining tree from the alignment and use this tree to infer the sequence's taxonomy.

   Treephyler follows a similar approach but uses speed-optimized Pfam domain prediction and treeing methods. Both approachesprovide more accurate classifications than those based on BLAST, but they work for fewer sequences, as Pfam hits are less frequent than BLAST hits (typically 20% of the genes).

18.8.2. Signature-based binning/classification

DNA base compositional asymmetries carry a weak but detectable phylogenetic signal that is most pronounced within the patterns of statistical over and under representation of tetra- to hexanucleotides. Various algorithms have been used to discriminate this signal from the DNA-compositional background noise and to use it for taxonomic inference. E.g. simple and advanced Markov models such as interpolated context models (ICMs), Bayesian classifiers and machine-learning algorithms such as support vector machines (SVMs), kernelized nearest-neighbor approaches and self-organizing maps (SOMs). Also, weighted PCA-based, Spearman distance-based and Markov Chain Monte Carlo-based assessments of oligomer counts have been used. As the information that DNA composition-based methods rely on is a function of sequence lengths, most of these methods deteriorate below 3–5 kb and perform poorly on sequences shorter than 1 kb.

Methods have been developed for successful binning of short reads as they are produced by Illumina machines, e.g. AbundanceBin, which is an unsupervised l-tuple-abundance-based clustering method. Recently, a signature-based method has been developed for fast taxonomic profiling of metagenomes that is independent of length and can be used with very short reads.

18.8.3. Mapping-based classification

Sequenced genomes have been used as references with known taxonomies for read recruitment in metagenome studies. This approach is particularly useful for habitats with species that have closely related sequenced relatives. A variant of this approach is to use habitat-specific sets of reference genomes for a competitive metagenome read mapping. Such sets can be compiled using the EnvO-lite environmental ontology. Low-quality repetitive reads should be excluded from the mapping using tools such as mreps and phages should be masked to minimize misclassifications. The mapping itself can be done with tools such as SSAHA2 or its successor SMALT (http://www.sanger.ac.uk/resources/software/smalt). Combined mapping information of the reads constituting a contig can be subsequently combined into a taxonomy consensus.

18.8.4. Combinatory classification

All aforementioned methods have specific advantages and disadvantages, and all are limited by the amount of information that can be retrieved from a sequence at all. Although protein-based methods tend to be more accurate than DNA-based methods, especially on shorter sequences or even reads, they can only classify sequences with existing homologues in public databases. But in metagenomics at least half of the genes of novel sequenced environmental microbes lack dedicated known functions, and a large proportion of these genes are hypothetical or conserved hypothetical genes that have either no or insufficient homologues with known taxonomic affiliation. This limitation does not apply to DNA-based methods, which, for their part, have other limitations. For example, methods such as ICMs, SVMs or SOMs need to be pre-trained, which is computationally expensive and must be continually done to keep pace with the fast-growing amount of new sequences.

In general, DNA-based methods suffer much more from a decrease in prediction accuracy when sequences get shorter than protein-based methods. Signature-based classifications hence work best with sequences from low to medium diverse habitats where ideally longer assemblies can be obtained or with habitats that feature species with a pronounced DNA composition bias.

Mapping-based classification is the most precise but is often hampered by the availability of suitable reference sequences to map to. As of this writing, 3171 genomes have been completed and 10 536 are ongoing according to the Genomes OnLine Database. Although this number seems impressive, entire clades of the microbial tree are poorly represented. The targeted sequencing of as yet unsequenced taxa like in the GEBA project and large-scale metagenome projects like the Earth Microbiome Project start to deliver large quantities of microbial genomes at an increasing pace. Hence, read mapping to closely related reference genomes might become the main method for metagenome taxonomic classifications in the not too distant future.

There is no standard for the taxonomic classification of metagenome sequences. Also, taxonomic sequence classification can be error prone, in particular for habitats with a complex diversity or high proportions of as yet barely characterized taxa. Rather than using a single method, acombination of individual methods is currently the most reasonable approach to partition metagenomes into taxobins. Such combinations have for example been implemented in PhymmBL that combinesICMs and BLAST and in CARMA3 that combines the original CARMA-approach with BLAST. In both cases the combination has already been shown to lead to increased classification accuracy. A combination of BLAST-, CARMA, SOM and 16S rRNA gene fragment-based classification termed 'Taxometer' was used in recent metagenome studies. Also, different binning methods have been successfully combined to improve accuracy. Besides combining different methods, it has recently been shown that combining multiple related metagenomes in a joint analysis is a way to improve binning accuracy.

One interesting aspect of taxonomic sequence classification is that it allows extrapolations onto relative taxon abundances. Although abundance information is lost in the assembly process due to the merger of similar sequences, abundance information can be obtained either from taxonomically classified reads or by back mapping of reads onto taxonomically classified assemblies. It has been shown that relative abundances obtained this way can be close to quantitative cell-based abundances assessments by CARD-FISH.

18.8.5. Pre-assembly taxonomic classification and binning

Binning and taxonomic classification methods are typically applied after the assembly. However, these methods can also be used prior to assembly to partition reads into taxonomic bins, which has the potential to substantially reduce the complexity of metagenome assemblies. This strategy might be particularly useful, when sequences from the habitat are already available (e.g. fosmids) that can serve as seeds in an iterative binning-assembly procedure.

18.9. Biodiversity estimation by16s rRNA gene analysis

About one in every few thousand genes in a metagenome data set is a 16S rRNA gene. With 454 pyrosequencing, typically translates to ~1000 reads per picotiter plate (~1 million reads) that harbor partial 16S rRNA genes with sufficient lengths and quality for phylogenetic analysis. Depending on the length and region of the retrieved partial 16S rRNA gene sequence, phylogenetic analysis can result into varying taxonomic depths. Since the introduction of 454+ a substantial fraction of the respective reads allows for a genus level assignment, and this situation is expected to even improve with future increases in pyrosequencing read length.

A limitation with pyrosequencing is that the number of obtained high-quality 16S rRNA genes might not be sufficient for representative biodiversity estimation, particularly not for lowly abundant taxa. Illumina does not have this problem due to its much higher throughput but on the other hand is plagued by its comparatively short reads that can compromise the depth and quality of the taxonomic assignments.

Representative sequences for operational taxonomic units (OTUs) can subsequently be mapped against a 16S rRNA reference tree for classification. The advantage of this method over 16S rRNA gene clone libraries is that no primers are involved and hence no primer bias exists. The disadvantage, besides not obtaining full-length high-quality 16S rRNA gene sequences, is that different taxa harbor different numbers of rRNA operons, which can distort metagenomic 16S rRNA gene abundances. For example, some Planctomycetes feature large genomes but only a single disjoint rRNA operon which would lead to an underestimation of their abundance in relation to average-sized genomes with more rRNA operons.

18.10. FUNCTIONAL ANALYSIS

Analysis of metagenomes involves functional annotation of the predicted genes by database comparison searches. This typically includes protein BLAST searches against databases such as SWISSPROT, NCBI nr or KEGG, HMMer searches against the Pfam [102] and TIGRfam databases, as well as predictions of tRNA and rRNA genes, signal peptides, transmembrane regions, CRISPR repeats and sub-cellular localization. Databases are available for special functions, for example the CAZY and dbCAN (http://csbl.bmb.uga.edu/dbCAN/) databases for carbohydrate-active enzymes, the TSdb and TCDB databases for transporters and the MetaBioMe database for enzymes with biotechnological prospects.

The resulting annotations are then used as a basis for functional data mining including metabolic reconstruction. Dedicated metagenome annotation systems have been developed to aid these tasks,e.g. WebMGA, IMG/M and MG-RAST. All three have expanded beyond mere annotation systems and continue to add useful features such as biodiversity analysis, taxonomic classification and metagenome comparisons. Tools have been developed for comparison, including METAREP, STAMP, CoMet and RAMMCAP.

METAREP and STAMP do not take sequence but already pre-processed data as inputs—tabulated annotations (such as gene ontology (GO) terms, enzyme commission numbers, Pfam hits and BLAST hits) in the case of METAREP and a contingency table of properties (for example exports from Metagenomics Rapid Annotation using Subsystems Technology (MG-RAST), Integrated Microbial Genomes (IMG)/M or CoMet) in the case of STAMP. Both tools feature various statistical tests and visualizations. METAREP is a web service developed by the J. Craig Venter Institute that can compare up to 20 or more metagenomes, whereas STAMP is a stand-alone software. The CoMet and RAMNCAP web servers in contrast do not require pre-computed data. CoMet takes sequence files as an input, does an Orphelia gene prediction, and subsequently runs HMMer against the Pfam database followed by multi-dimensional scaling and hierarchical clustering analysis on the Pfam hits and associated GO terms plus visualization of the data. RAMMCAP takes raw reads as an input, does a six-reading frame open reading frame (ORF) prediction, clusters reads and ORFs, does a HMMer and BLAST-based annotation and allows comparison of the data, e.g. by similarity matrices. RAMMCAP is part of the CAMERA data portal, which currently comes closest to an integrative processing pipeline for metagenomes with various tools for data retrieval, upload, querying and analysis.

Although automatic in silico annotation is essential for metagenome analysis, one should not forget that a substantial proportion of such annotations are erroneous or even incorrect. Aside from well-studied pathways of the core metabolism, automatic annotations are also often unspecific, i.e. restricted to assigning general functions (e.g. lipase, oxidoreductase, alcohol dehydrogenase) without resolving the involved specific substrates and products. This reflects a fundamental lack of knowledge rather than a limitation of bioinformatic methods per se and can only be addressed by future high-throughput functional screening pipelines. One of the intriguing aspects of metagenomics is that typically about half of the genes in a metagenome have as yet unknown functions. Hence, restricting metagenome analyses to genes with functional annotations equals to ignoring large proportions of the genes. As a solution, it has been proposed to cluster and analyze metagenomic ORFs in a similar way as OTUs in biodiversity analyses. Such clusters have been termed operational protein families and can be analyzed, for example with MG-DOTUR.

18.11. AUTOMATIZATION, STANDARDIZATION AND CONTEXTUAL DATA

Until recently, the capacity to sequence has been the limiting factor for metagenome analysis. The continual increase in sequencing capacity and decline of costs mean while have turned postmetagenomic data analysis into the main bottleneck. Metagenome projects tend to suffer from insufficient resource allocation for data post-processing. Steps in metagenome analyses that can be automated should be automated to ensure quality, but this requires the establishment of commonly accepted data formats for metagenome sequences and their associated contextual (meta)data, as well as defined interfaces for data exchange and integration—a task that is tackled by the Genomic Standards Consortium (GSC, http://gensc.org).

Contextual data are among the key factors for successful metagenomes analyses, in particular when it comes to interpretation of time series or biogeographic data. Contextual data are all the data that are associated with a metagenome, such as habitat description (including geographic location and common physicochemical parameters) and sampling procedure (including sampling time). The GSC has published standards for the minimum information about a metagenome sequence (MIMS) as part of the minimum information about any sequence (MIxS) standards and checklist, which are supported by the International Nucleotide

Sequence Databases Collaboration (INSDC). Similarly, standards have been devised in terms of data formats to ensure data inter-operability, such as the genomic contextual data markup language. It is important that contextual data are collected and integrated into databases, because in the long run these data will allow to extract correlations between geography, time, prevailing environmental conditions and functions from metagenomic data.

As of today, there is no comprehensive tool for metagenome analysis that incorporates all types of analysis (biodiversity analysis, taxobinning, functional annotation, metabolic reconstruction and sophisticated statistical comparisons). In terms of pipelines, MetaAMOS provides an integrated solution for the initial post-processing of mated read metagenome data that supports different assemblers, the BAMBUS 2 scaffolder and various gene prediction, annotation and taxonomic classification tools. In terms of data integration, CAMERA has so far developed the most comprehensive infrastructure for holistic metagenome analyses, and further tools and pipelines are currently developed in the GSC and Micro B3 project (http://www.microb3.eu/) frameworks.

18.12. DATA SUBMISSION

Progress in sequencing allows for metagenomes with increasing sizes. A full run on an Illumina HiSeq2000 sequencer does not only produce 600 Gb of sequences but also the FASTQ raw data files are multiple times as large. Although sequencing facilities send these data to their customers on Terabyte-scale hard disc drives, such data volumesare certainly not suitable anymore for upload to data analysis servers or INSDC databases for submission, even not with fast user datagram protocol (UDP) protocols such as Aspera Connect. This is a problem that is as yet unsolved. Also, the INSDC databases are currently not prepared for handling the quality of many metagenomes (pervasiveness of frameshifts, automatically generated non-standard annotations and large amounts of partial genes) and their accessory data (such as lists of metagenomic 16S rRNA gene fragments including taxonomic classifications). It is clear that currently sequencing technologies evolve faster than bioinformatic infrastructures for post-genomic analysis are built. As mentioned before, this has been recognized and efforts such as those of the GSC, Micro B3, CAMERA, MG-RAST and IMG/M are on the way to define standards and develop pipelines for future metagenome data handling. The European Bioinformatics Institute has made a submission and analysis pipeline for 454/ pyrosequencing metagenome data available (https:// www .ebi .ac. uk/ metagenomics).

18.13. FUTURE PERSPECTIVES

The newest generation of sequencers, such as the PacBio RS, the Ion Torrent Proton or the ONT GRIDIon/MINIon, will continue to propel the field of metagenomics, and who knows whether at some point in the future technologies such as the conceptual IBM/Roche DNA transistor will revolutionize the field again. On the one hand, development of affordable bench-top devices (454 Junior, Illumina MiSeq, Ion Torrent PGM and Proton) has led to a democratization of sequencing, and future devices such as the ONT MINIon could even be used for metagenomic analyses directly in the field. On the other hand, the ever-growing throughput of NGS sequencers is making data analysis increasingly complex. Although smaller and medium-sized metagenomes can be analyzed with the resources described so far, different infrastructures and bioinformatic pipelines are necessary for future large-scale projects. 'Megagenome' projects reach the size of many terabytes of sequences (and beyond), and instead of moving these data around, it is reasonable that they reside at the sequencing institution and that these institutions provide pipelines for remote data analyses. This implies that large-scale sequencing and large-scale computing have become inseparable. For example, the BGI (formerly Beijing Genomics Institute, http://en.genomics.cn) has projected an integrated national center for sample storage, sequencing, data storage and analysis. Monolithic data centers are one way to address this, but also cloud computing such as Amazon's EC2/S3 can be used as a viable and scalable alternative for large-scale metagenome data analysis provided that the data can be transferred to the cloud, and appropriate data security is guaranteed.

Metagenomics constitutes an invaluable tool for investigating complete microbial communities *in situ*, in particular when integrated with biodiversity, expression and contextual data (metadata). Continuous advancements in sequencing technologies not only allow for addressing more and more complex habitats but also impose growing demands on bioinformatic data post-processing. Not long ago, sampling and associated logistics, clone library construction

Key Points to remember

- Metagenomics has become an indispensable and widely affordable tool for studying as yet uncultivable microbes (Bacteria,Archaea and viruses).
- Progress in NGS allows for larger metagenomes, for studying series of metagenomes over time and space and for addressing increasingly complex habitats.
- A new type of integrative ecosystems biology study seeks to combine metagenomics with metatranscriptome, metaproteome, metabolome and biodiversity and contextual (meta)data analyses.

- There are several bioinformatic tools and pipelines for different aspects of metagenome analysis, but there is no standardized,comprehensive pipeline covering all aspects. Large-scale 'megagenome' projects are particularly affected and hence face challenges with respect to data handling, data integration and data analysis.
- Ongoing international efforts strive to establish standards and tools for future large-scalemetagenome analysis that are necessary to turn the proverbial metagenomic data deluge into knowledge.

Mini Quiz

1. What is the relationship between biodiversity and bioinformatics?
2. Name the platforms which offer longer read lengths?
3. Define pseudo-replication?
4. Explain the methods for successful binning of short reads?

CHAPTER - 19

NANOTECHNOLOGY IN METAGENOMICS

Prokaryotes are the most numerous group of living organisms numbering approximately 10^9 cells per gram of soil and up to 8300 000 different species. They represent a tremendous reservoir of genetic information and diversity and their screening for useful properties is frequently a first step in the discovery of useful molecules, genes, proteins etc. by bio-industry. Traditionally, this screening has relied upon conventional microbial cultivation and isolation techniques, but these methods only permit the isolation of a small fraction of the micro-organisms present, because many are viable but uncultivable. In this context, the isolation of environmental metagenomes (i.e. the genomes of the total microbiota present in a specific environment) without specific reference to particular micro-organisms has been recognized as an attractive and powerful approach to mine novel genetic resources in the natural environment.

The efficient isolation of nucleic acids, especially DNA from soil samples, could provide a method to access the entire soil metagenome. Currently, methods for the extraction of DNA from soil samples generally rely on the use of phenol D chloroform but the process is toxic, time-consuming, and multi-step and utilizes organic solvent extraction, alcohol precipitation, as well as centrifugation. The method can also be inconvenient or impossible to utilize when dealing with either small amounts of DNA or large numbers of samples.

19.1. NANOPARTICLES IN BIOSEPERATION OF NUCLEIC ACIDS

In the context of bioseparation and purification, magnetic carrier technology has become an increasingly popular tool for the separation of biomolecules (e.g.

DNA, RNA and proteins). (Magnani *et al.* 2006) and this group has previously reported the fabrication of super paramagnetic silica-magnetite nanoparticles and their use in the extraction of nucleic acids from bacterial cells. The isolation of DNA from soil samples based on the use of silica-magnetite nanoparticles, is a rapid, inexpensive and scalable method. The DNA obtained is of high molecular weight and suitable for subsequent molecular biological manipulations, such as PCR and endonuclease restriction digestion.

19.2. MAGNETIC BIOSEPARATION

- Resuspend sieved soil (1.5 g) in 1 ml lysis buffer [100 mmol l^{-1} Tris–HCl pH 8.0, 100 mmol l^{-1} EDTA pH 8.0, 1 mol l^{-1} NaCl, 2% wD v sodium dodecyl sulfate (SDS), 100 μg ml^{-1} RNase A in a 15-ml Falcon tube
- Incubate at 60°C for 30 min with end-over-end rotation.
- Centrifuge the sample (10 000 g for 5 min) and transfer the supernatant to a clean 2 ml Eppendorf tube containing 1 mg of silicamagnetite nanoparticles.
- Add same volume of binding buffer [20% wD v polyethylene glycol (mol wt 8000) in 4 mol l^{-1} NaCl] and the mixture incubated for 5 min at room temperature with end-over-end rotation.
- Subsequently immobilize the silica-magnetite nanoparticles using a magnetic stand
- Discard the supernatant and wash the nanoparticles three times by adding 200μl of 70% v D v aqueous ethanol and incubate for 2 min at room temperature with endover-end rotation.
- Elute the DNA from the particles by adding 100μl nuclease free water and incubate the suspension for 5 min at room temperature with an end-over-end rotation (maximum DNA recovery obtained when the water pH is between pH 7.0 and pH 8.5).
- Immobilize the nanoparticles and transfer the DNA-containing supernatant to a sterile 1.5 ml Eppendorf tube.
- The elution process has to be repeated twice and combined.
- Store the resulting DNA at 20°C until required.

19.3. NANOPORE SEQUENCING – THE FOURTH GENERATION SEQUENCING

The nanopore technologies achieve sequencing by single-molecule sequencing without amplification, real-time sequencing without repeated cycles and synthesis can be eliminated. This is the reason that it is classified as 4G sequencing technology. Nanopore sequencing is based on the concept that single molecules of DNA could be identified by passing through a tiny channel.

The concept of using a nanopore as a biosensor was first proposed in the mid 1990s when research into nanopores was beginning at academic institutions such as Oxford, Harvard and UCSC. Oxford Nanopore was founded in 2005 to translate academic nanopore research into a commercial, electronics-based sensing technology. The comprehensive end-to-end system includes sample preparation, molecular analysis and informatics, and is designed to provide novel benefits to a range of users for a broad number of applications.

Table 19.1 Comparison of DNA extraction method from soil samples

Parameters	Methods		
	Magnetic bioseparation	**Phenol D chloroform***	**Soil Master DNA extraction kit**
A260 D A280	0.98 ± 0.03	1.09 ± 0.04	1Æ58 ± 0Æ37
A260 D A230	0.5 ± 0.03	0.78 ± 0.03	0.57 ± 0.28
DNA yield (μg)	9.37 ± 2.3	4.62 ± 0.67	0.60 ± 0.12
No. of samples	3	3	3
Time required	2 h	5 h	2 h
Cost per sample	1.5	3.0	8.0
Method	Easy	Difficult	Easy

19.3.1. Nanopore

A nanopore is simply a small hole, of the order of 1 nanometer in internal diameter. Certain porous transmembrane cellular proteins act as nanopores, and nanopores have also been made by etching a somewhat larger hole (several tens of nanometers) in a piece of silicon, and then gradually filling it in using ion-beam sculpting methods which results in a much smaller diameter hole: the nanopore.

19.3.2. Principle

The theory behind nanopore sequencing is that when a nanopore is immersed in a conducting fluid and a potential (voltage) is applied across it, an electric current due to conduction of ions through the nanopore can be observed. The amount of current is very sensitive to the size and shape of the nanopore. If single nucleotides (bases), strands of DNA or other molecules pass through or near the nanopore, this can create a characteristic change in the magnitude of the current through the nanopore.

19.3.3. Nanopore fabrication

A nanopore is, essentially, a nano-scale hole. This hole may be:

- **biological:** formed by a pore-forming protein in a membrane such as a lipid bilayer
- **solid-state:** formed in synthetic materials such as silicon nitride or graphene
- **hybrid:** formed by a pore-forming protein set in synthetic material

19.3.3.1. *Alpha hemolysin*

Alpha hemolysin (áHL), a nanopore from bacteria that causes lysis of red blood cells. To All the four bases can be identified using ionic current measured across the áHL pore. The structure of áHL is advantageous to identify specific bases moving through the pore. The áHL pore is ~10 nm long, with two distinct 5 nm sections. The upper section consists of a larger, vestibule-like structure and the lower section consists of three possible recognition sites (R1, R2, R3), and is able to discriminate between each base. Sequencing using áHL has been developed through basic study and structural mutations, moving towards the sequencing of very long reads. Protein mutation of áHL has improved the detection abilities of the pores] The next step is to bind an exonuclease onto the áHL pore. The enzyme would periodically cleave single bases, enabling the pore to identify successive bases. Coupling an exonuclease to the biological pore would slow the translocation of the DNA through the pore, and increase the accuracy of data acquisition.

19.3.3.2. *MspA*

Mycobacterium smegmatis porin A (MspA) is the second biological nanopore currently being investigated for DNA sequencing. The MspA pore has been identified as a potential improvement over áHL due to a more favorable structure. The pore is described as a goblet with a thick rim and a diameter of 1.2 nm at the bottom of the pore. A natural MspA, has a negative core that prohibited single stranded DNA(ssDNA) translocation. The natural nanopore was modified to improve translocation by replacing three negatively charged aspartic acids with neutral asparagines. The electric current detection of nucleotides across the membrane has been shown to be tenfold more specific than áHL for identifying bases. The dsDNA would halt the base in the correct section of the pore and enable identification of the nucleotide.

19.3.3.3. *Fluorescence*

An effective technique to determine a DNA sequence has been developed using solid state nanopores and fluorescence. This fluorescence sequencing method converts each base into a characteristic representation of multiple nucleotides which bind to a fluorescent probe strand-forming dsDNA. With the two color system proposed, each base is identified by two separate fluorescences, and will therefore be converted into two specific sequences. Probes consist of a fluorophore and quencher at the start and end of each sequence, respectively. Each fluorophore will be extinguished by the quencher at the end of the preceding sequence. When the dsDNA is translocating through a solid state nanopore, the probe strand will be stripped off, and the upstream fluorophore will fluoresce. This sequencing method has a capacity of 50-250 bases per second per pore, and a four color fluorophore system (each base could be converted to one sequence instead of two), will sequence over than 500 bases per second. Advantages of this method are based on the clear sequencing readout using a camera instead of noisy current methods. However, the method does require sample preparation to convert each base into an expanded binary code before sequencing. Instead of one base being identified as it translocates through the pore, ~12 bases are required to find the sequence of one base.

19.3.3.4. *Nanopore sensing*

A nanopore may be used to identify a target analyte as follows:

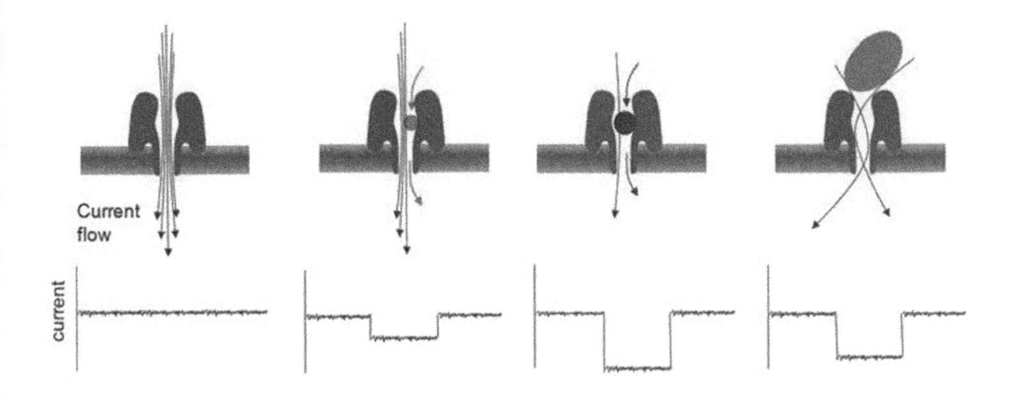

19.4. CHALLENGES IN NANOPORE SEQUENCING TECHNOLOGY

Though the nanopore sequencing offers many advantages the challenges remain in its resolution to detect single bases and integration of exonuclease in the nanopore detection systems. Replacing protein pores with solid-state channels could increase the advantages of nanosequensers.

In this field, IBM and Roche are collaborating to explore electronics-based approaches for the development of a 'DNA transistor' as nanopore-sequencing technology. Graphene pore is another solid-state pore material, enabling a single base to pass by at one time improving the accuracy of sequencing and detection.

19.5. CONCLUSION

In conclusion, the nanopore field is emerging with gusto. On the curvy road towards DNA sequencing, nanopores have sparked the imagination of scientists as tools that can help in solving problems in biophysics. Similar to atomic force microscopy, optical/magnetic tweezers, and fluorescence microscopy, nanopores are taking stage as one of a handful of main tools for exploring individual protein and DNA molecules including metagenome.

Mini Quiz

1. What is the role of nanoparticles in nucleic acid separation?
2. Explain the principle of nanopore sequencing?
3. Name the nanopore used for DNA sequencing?
4. What are the merits of nanopore sequencing over pyrosequencing?

CHAPTER - 20

THE FUTURE OF METAGENOMICS

Metagenomics is one of the fastest advancing fields in biology. By permitting access to the genomes of entire communities of bacteria, viruses and fungi otherwise inaccessible, metagenomics is extending our comprehension of the diversity, ecology, evolution and functioning of the microbial world, as well as contributing to the emergence of new applications in many different areas. The continual and dynamical development of faster sequencing techniques, together with the advancement of methods to cope with the exponentially increasing amount of data generated, are expanding our capacity for the analysis of microbial communities from an unlimited variety of habitats and environments. The synergism with the new emerging "omics" approaches is showing the path to functional metagenomics and to adopting integrative, wider viewpoints like systems biology.

20.1. METAGENOMICS FOR A SUSTAINABLE WORLD

In recent years several industrial processes and products have become more environmentally friendly by replacing chemical processes with microbial processes.

In addition, microbial processes offer new methods for cleaning water and soil. Genomics can play an important role in speeding the transition for a fossil fuel to a bio-based society by offering tools for the characterization of microorganisms and the discovery and selection of new genes that form the blueprint for new enzymatic pathways. The questions that arise are how industry can use genomics to address societal pressure to manufacture products both cost-effectively and environmentally clean. The possibilities seem enormous, but to explore and then exploit these possibilities, new means of cooperation between universities and industry have to be developed and a 'trialogue' will have to take place among science, industry and society.

20.2. METAGENOMICS HOLDS PROMISES FOR PEOPLE, PLANET AND PROFIT

Sustainability pertains to society, ecology and economy,or as indicated in business terms : People, Planet and Profit. Genomics holds promise for all three Ps.

'Try to get a better world for the people who live on it, with sustainability as the motor for development and innovation.' This statement was given by Van Eijndhoven, president of the Executive Board of Erasmus, University in Rotterdam during Genomics Momentum, 2004.

Technology is not something that develops on its own and is meekly followed by society. It is interplay of both parties, emphasized Van Eijndhoven. Social acceptance of genomics is therefore an important issue.The public is still quite skeptical following the sudden introduction of improved agricultural products with genetic technology. 'The idea that once people get used to it, they will accept it is not always true. Radiation of food, for example, never gained acceptance. The benefits of early applications of green biotechnology in agriculture were only clear to the industry, and not to consumers. On the other hand, people forget that the production of insulin for diabetes is also the result of genetic modification.'

While metagenomics and genetic modification are in the same research area, they are different. Genes don't have to be modified in order to get things done. In that sense, genomics is regarded as less 'risky.' But even zero risk is not good enough if there are no benefits to be gained. People's perception of the resulting profits and risks can have a profound influence on the future of genomics.

Metagenomics will be part of the good life in the 21st century, according to Van Eijndhoven. In order to make this happen we have to maintain a good sustainable and trustworthy relationship with society inform people about what's happening and interact with the stakeholders on a continuous basis to gain and keep society 's license to produce and develop.

20.3. MINING GENOMES FOR NOVEL ENZYMES

Microorganisms are the most important employees for the production of fine chemicals, food ingredients and pharmaceuticals. Several genome sequences are already available, for organisms such as *Aspergillus niger* which is a source of new enzymes, *Penicillium chrysogenum* for antibiotics, *Saccharomyces cerevisiae* for live cultures and enzymes, and *Escherichia coli* for fine chemicals and enzymes. With genomics it is possible to make strategic choices in strain and process development for the production of new enzymes. At first, 126 genes of the *A.niger* genome were known, representing only 0.9 percent of the total of 14,078 genes. DSM, a fermentation company in USA now has a functional prediction for 7428 genes, of which half are required for the cell's metabolic organization. With this in mind, *A.niger* offers many leads for strain and process improvement and for many potential new products. It is now possible, to compare bad and good properties of different strains and processes very easily by using an Affymetrix Gene Chip. Differences in mRNA, metabolite and protein content can now be related to distinct properties. Such array analysis is therefore now being implemented in strain and process improvement projects.

By analyzing the complete DNA sequence of the *Aspergillus niger* fungus,the DSM company gained access to a whole new variety of previously unknown enzymes such as proteases and peptidases. About 200 genes encode for proteases, of which about 60 are secreted. Combinations of proteases can again lead to many different peptides originating from many different sources of protein such as rice, whey, maize and casein., These peptides can find applications in sport nutrition, diabetes, weight management and infant formula. One of these enzymes is now used for the preparation of a specific peptide mixture, which finds application into a new muscle recovery drink for athletes. The proteolytic enzyme degrades proteins in specific fragments .The sports drink is now known under the name PeptoPro for faster muscle refueling. After physical training or a severe contest, the energy reserve battery of the body is almost empty. The difference between DSM 's sports drink and any other energy drink is that the energy battery is recharged more quickly and a complete recovery is reached. The drink increases the capacity of muscles to absorb glucose and has been found to increase the average performance level of athletes.

Another enzyme works by masking the bitter taste of hydrolyzed protein materials, a well-known handicap of peptide mixtures, which often limits or even prohibits the use of this valuable and healthy food ingredient. This will increase the application of milk and whey protein material. The novel asparignase from A. niger is another example. The enzyme can prevent acrylamide formation in baked products by converting the amino acid asparagine into aspartate.

Genomic information and tools are of paramount importance to speed up strain and process development and the discovery of novel enzymes. In a short time span, DSM has discovered and developed several new enzymes. But the development of high-throughput application screens is needed to keep up. 'Finding the genes that do it is like finding needles in haystacks, but it is becoming a core business. And if there are no haystacks to explore we will create them ourselves by directed evolution,' a quote by Van Leen, DSM food specialist.

20.4. METAGENOMICS FOR A SUSTAINABLE BIO-BASED COMPANY

A bio-based economy that uses renewable raw materials to produce energy and products, offers sustainable solutions, to reduce the production of greenhouse gases, to improve the quality of air and water and to provide food for all people. Such an economy is needed, because conventional production methods based on chemistry and catalysis are depleting the fossil fuels and generate a great deal of waste, which ends up in landfills or incinerators But a bio-based economy, uses renewable sources like cornstarch and bio-catalytic processes performed by enzymes, arrays of enzymes or whole cells, wastes or byproducts are recycled and used again.

A bio-based economy is also sustainable in an economic sense. Studies on the impact of biotechnology predict that as soon as 2010,up to 10 or 20 percent of the chemical sales would be based on bioprocesses. But the broad implementation will depend largely on policy framework, demand, feedstock prices and, ofcourse, investments. For this low-cost carbon is needed, either from traditional agricultural feed stocks (corn) or biomass feedstock (plant waste).A lot of these low-cost carbon sources are difficult to use. Leftovers from corn plants, for instance, mainly consist of long rigid cellulose molecules that are tough to digest. Still , they are made of sugar molecules and since sugar is the basis for many bioprocesses, the challenge is to convert cellulose in a cost-effective manner into fermentable sugars with the aid of enzymes. Bioconversion of sugar provides industry with a large number of building blocks that form the basis for many products, ranging from food supply and pharmaceutical to paints and plastics. To this end, new technological breakthroughs in enzyme-based processing of raw plant material are still needed to make the way to low-cost carbon.

To enable and accelerate the transformation to a bio-based economy, Genencor International developed a microbial strain for the synthesis of 1.3 propanediol, which is the basic component for the production of a new type of plastic. Since no single microorganism is able to do the necessary transformation from sugar to 1.3 propanediol on its own, the metabolic pathway had to be designed. Genes were recruited from three different microorganisms. The host and production strain *Escherchia coli* (*E.coli*) was provided with a pathway of baker's yeast to make the intermediate glycerol and complemented with a pathway of the bacteria *Klebsiella* to make the final product 1,3 propanediol. In total, eight new genes were introduced in the plasmid. Eighteen genes of the host *E.coli* were modified to make the new metabolic pathway work, such as adapting the transport systems in order to get substrate in and product out of the cell and delivery of the reducing power for conversion steps. On basis of this constructed metabolic pathway system for 1.3 propanediol, the company is in an excellent position to develop other cell factories to make new components. The same pathway design can be used to make products, such as antifreeze, tanning agents, antimicrobial agents and glycerol. This will make it an interesting competitor to chemical processes for making small molecules in large volumes, and finally enhancing broad implementation when low-cost carbon is available. With the desired product in mind and with input from biochemical analysis, transcriptomics, and proteomics, the algorithm estimates all options, and chooses the best possibilities.

20.5. ECO AND EVOLUTIONARY GENOMICS

At first sight, the disciplines of ecological and evolutionary genomics might seem strictly fundamental and of no direct use to society. Today species are becoming extinct at an unprecedented rate, even before their existence or their special properties were known. Genomics can also help increase understanding of evolutionary processes, such as the emergence of new infectious diseases and antibiotic resistance, and is therefore of direct significance for the wellbeing of mankind. Since a few genes can make the difference between extinction and survival, genomics can help to identify the genes that really matter in evolution and ecology .Using clues found in the repeatability of evolution in large experimental populations, might even help to predict which types of infectious variants of micro-organisms might evolve.

The genetic diversity of the microbial world is even more than expected. Metagenomics of the ocean begins to reveal the treasures of the microbial world. While the databank of sequences known to man, including those from the human genome, contained more than 180,000 genes, sequencing genomic information in 1,500 litres of Sargasso seawater revealed two million new ones. No doubt that many useful products are hidden in this diversity. With this prospect in consideration its good to remember that major breakthroughs in research have been based on curiosity and individual excellence. The thermostable enzyme that is so crucial to PCR was discovered during research of micro-organisms in a hot spring in Yellow Stone National Park in the US. So in order to develop real innovations and make real progress, curiosity-driven research has to be preserved. Then ecological and evolutionary genomics can contribute to a more sustainable world to their full extent.

20.6. METAGENOMICS FOR SUSTAINABLE INDUSTRIAL PRODUCTION

Metagenomics promises to improve industrial production by optimizing microbial processes and replacing chemical productions steps by simple microbial processes White, or industrial ,biotechnology is a viable and sustainable concept for improving large-scale production of fuel and bulk chemicals such as plastics and fibres.

For sustainable production of fine chemicals, novel microbial strains must and are being developed. Hereby it is possible to shift from organic synthesis toward single-step microbial processes while delivering economic as well as environmental benefits. Genomics can be used as a tool to improve these existing micro-organisms by metabolic engineering or in a classical manner by mutagenesis and selection. On the other hand, genomics can help industrial production processes by optimizing the process conditions to reveal the behaviour of different production strains on a molecular basis. The combination of both approaches has already led to a production method for biofuels from wood-based materials that would otherwise be disposed or incinerated, for instance. Using genetically modified micro-organisms introduces an additional risk for humans and the environment. Because of this area of tension, the industry is searching for a license to operate, environmental groups, weigh the promise of genomics and industrial microbial production against potential risks and alternative methods for making processes more sustainable. A dialogue between industry and non-governmental organizations (NGOs) is needed for its success.

20.7. CARBON CYCLING AND CLIMATE

The global carbon cycle plays a central role in regulating atmospheric carbon dioxide levels and thus Earth's climate, but our basic understanding of the tightly interlinked biological processes driving the carbon cycle remains limited. Advancing our knowledge of these processes is crucial to predicting potential climate change impacts, assessing the viability of climate change adaptation and mitigation strategies, and informing relevant policy decisions

20.8. CONCLUSION

The increasing global demand for energy, materials and goods draws on a finite and diminishing reserve of fossil fuels. To meet future needs, alternative renewable sources of energy and bio-based feedstocks must be developed. Genomics and industrial microbial fermentation offer the tools for such a sustainable development and the transition to a bio-based economy and society. Broad implementation of these sustainable techniques will depend heavily on policy framework, demand, feedstock prices, research investments and societal embedding.

Companies in Europe are mining genomes for new and sustainable applications, aided by a large set of new tools micro-arrays, functional genomics, structural genomics, proteomics and bioinformatics created in recent years. To keep up and find the right genes first,development of new techniques,high- throughput applications and new technological breakthroughs are still needed. With these elements in place,the outcome of research and the possibilities for a sustainable world are still beyond imagination. To keep up in thsis economic battle field, a dialogue between the public and policymakers is more needed than ever. The potential of genomics and industrial microbial production are too promising to be bogged down by miscommunication and mistrust. Societal acceptance of genomics is therefore a crucial ingredient to success. If this barrier is solved then genomics will become an integrated part of our culture and of the sustainable world of the 21 st century.

Imagine! A future in which we can

- use "super bugs" to detect chemical contamination in soil, air, and water and clean up oil spills and chemicals in landfills;
- cook and heat with natural gas collected from a backyard septic tank or bottled at a local waste-treatment facility;
- obtain affordable alcohol-based fuels and solvents from cornstalks, wood chips, and other plant by-products; and
- produce new classes of antibiotics and process food and chemicals more efficiently.

Mini Quiz

1. What is the future of metagenomics?
2. Elucidate the recent advents of metagenomics in sustainable industrial production?
3. How does metagenomics impact climate change?

GLOSSARY

***5-Bromo-2-deoxyuridine labeling*:** BrdUTP labelling offers an alternative in cases where SIP labelled compounds are not available. Growth in the presence of BrdUTP and the unlabeled compound accesses metabolically active organisms. These methods are limited by the difficulties in acquiring high labelling efficiency and the recycling of the label in the

***Abiotic*:** Non-living objects, substances or processes.

Accession number: an identifier associated with a sequence or entity supplied by a particular biological database that uniquely identifies the given sequence or entity within that database.

***Affinity capture*:** Oligonucleotides covalently immobilised to a solid support can be used to affinity purify target genes. The slow kinetics of hybridization limit this process, but might be improved by using metagenomics mRNA or single-stranded DNA.

***Alignment*:** The process of lining up two or more gene or protein sequences to achieve maximal levels of identity (and conservation, in the case of amino acid sequences) for the purpose of assessing the degree of similarity and the possibility of homology.

***Annotation*:** to comment on the genomic consensus sequence. Specifically, identifying genes and determining their function; adding pertinent information such as gene coded for or coding sequence, amino acid sequence, or other commentary to the database entry of raw DNA sequence.

***Assembly*:** the process of reconstructing larger genomic sequences from smaller randomly derived subsequences. This process relies on determining sequence similarity between overlapping sequences. Also refers to the collection of assembled sequences that are the output of an assembly program.

***Bacterial artificial chromosome* (BAC):** an artificially constructed vector for medium-sized segments of DNA (up to 300 kb in length), which are then incorporated into a host cell (usually *Escherichia coli*). BACs serve as cloning vectors in metagenomics.

***Bioinformatics*:** the science of managing and analyzing biological data using advanced computing techniques.

***Biopanning*:** It involves repeated cycles of binding that will successively enrich the pool. After several rounds of enrichment, individual clones are characterised by DNA sequencing. This method is efficient and amenable to high-throughput screening, offering the potential to enrich even rare DNA sequences in the metagenome, but current phage technology limits expression of proteins !50kDa.

***BLAST* (Basic Local Alignment Search Tool):** A sequence comparison algorithm optimized for speed used to search sequence databases for optimal local alignments to a query. The initial search is done for a word of length "W" that scores at least "T" when compared to the query using a substitution matrix. Word hits are then extended in either direction in an attempt to generate an alignment with a score exceeding the threshold of "S". The "T" parameter dictates the speed and sensitivity of the search.

***Clear Range*:** the portion of a sequencing read that contains high quality sequence data. By definition this excludes vector sequence and low quality error prone sequence.

Cloning vector: a small DNA vehicle that can accommodate a foreign (cloned) DNA fragment. Plasmids, cosmids, fosmids and BACs are examples of cloning vectors.

***Coding Sequence* (CDS):** that portion of a gene or mRNA which directly specifies the amino acid sequence of a predicted protein.

***Conservation*:** Changes at a specific position of an amino acid or (less commonly, DNA) sequence that preserve the physico-chemical properties of the original residue.

***Consortium*:** Physical association between cells of two or more types of microorganisms. Such an association might be advantageous to at least one of the microorganisms.

***Contig*:** a contiguous length of assembled consensus genomic sequence in which the order of bases is known to a high confidence level.

***Cosmid*:** hybrid plasmid constructed by the insertion of cos sequences, which are DNA sequences of the bacteriophage Lambda.

***Differential expression analysis* (DEA):** DEA targets transcriptional differences in gene expression. Several variations in the basic concept exist. These include selective amplification via biotin and restriction-mediated enrichment (SABRE), integrated procedure for gene identification (IPGI), serial analysis of gene expression (SAGE), tandem arrayed ligation of expressed sequence tags (TALEST) and total gene expression analysis (TOGA).

***Diversity*:** statistically, it describes the richness as well as distribution of relative abundance of a species compared to the different species found in a sample. Microbial diversity in soil is likely to be the highest found in any ecosystem on Earth.

***E value* (Expectation value):** the number of different alignments with scores equivalent to or better than S that are expected to occur in a database search by chance. The lower the E value, the more significant the score.

***Filtering*:** Also known as masking. The process of hiding low-complexity regions of (nucleic acid or amino acid) sequence that frequently leads to spurious high scores.

Fluorescence-activated cell sorting (FACS): sorting of cells that have been tagged with a fluorescent dye in a flow cytometer.

***Fosmid*:** a hybrid vector consisting of an f-factor cosmid (circular DNA) that is capable of containing larger pieces of DNA, i.e. up to 50 kb (average 35 kb) compared to only 10 kb in a plasmid.

***Functional Annotation*:** the functional characterization and classification of a sequence (typically protein coding). This typically consists of describing the biological function, enzymatic function, and localization using a variety of pre-defined terms or identifiers.

***Gap*:** a space introduced into an alignment to compensate for insertions and deletions in one sequence relative to another. To prevent the accumulation of too many gaps in an alignment, introduction of a gap causes the deduction of a fixed amount (the gap score) from the alignment score. Extension of the gap to encompass additional nucleotides or amino acid is also penalized in the scoring of an alignment.

Homology: related through descent from a common ancestor.

Host: in cloning procedures, hosts are organisms that serve as the recipient of the cloning vectors that each carry a unique copy of foreign DNA directly extracted from the environment or from another organism. The most common metagenomic library host is the bacterium *E. coli*.

Humus: A complex of heteropolymeric substances, including humic acids, humin and fulvic acids.

Identity: The extent to which two (nucleotide or amino acid) sequences are invariant.

Low Complexity Region: regions of biased composition including homopolymeric runs, short-period repeats, and more subtle overrepresentation of one or a few residues.

Mate pair: refers to the set of two reads resulting from pairwise end sequencing

***Metabolite-regulated expression* (METREX):** technique using a reporter plasmid in the host organism which is directly regulated by compounds that induce quorum sensing. Once a threshold of the inducer is produced within the cell, GFP will be expressed.

Metagenomics: the study of the collective genomes recovered from environmental samples without prior cultivation. It enables the investigation of genome information on organisms that are not easily cultured in the laboratory. It is therefore a means of systematically investigating, classifying and manipulatingthe entire genetic material isolated from environmental samples.

Microarray: Microarrays allow high-throughput robotic screening for targeting multiple gene products. The cost and availability of microarray technology is rapidly decreasing, making this an increasingly attractive option.

Next generation sequencing (NGS) technology: Roche 454, Illumina, or Life Technologies SOLiD allows for sequencing without the need for cloning DNA fragments into *E. coli* . NGS technologies are more affordable; provide greater reads, more depth of coverage, and lower error rates compared with Sanger-based methods but the read lengths have been limited to short reads of 35 to 250bp until recently. NGS sequencing typically requires PCR with barcoded primers, yielding amplified products that can be separated and categorized following sequencing.

Open Reading Frame (ORF): series of codons or nucleotide triplets without any termination codons. There are six potential reading frames of an unidentified nucleotide sequence and these are (potentially) translatable into a protein.

Orphelia: It is a metagenomic ORF finding tool for the prediction of protein coding genes in short, environmental DNA sequences with unknown phylogenetic origin. Orphelia is based on a two-stage machine learning approach that was recently introduced by our group. After the initial extraction of ORFs, linear discriminants are used to extract features from those ORFs. Subsequently, an artificial neural network combines the features and computes a gene probability for each ORF in a fragment. A greedy strategy computes a likely combination of high scoring ORFs with an overlap constraint.

Phage display**:** Phage-display expression libraries provide a means of isolating a given DNA sequence by affinity selection of the surface-displayed protein to an immobilised ligand.

Plasmids**:** independent, circular and self-replicating DNA molecules that carry only a few genes. Plasmids are autonomous molecules and exist in cells asextra chromosomal genomes, although some plasmids can be inserted into a bacterial chromosome.

***Product induced gene expression* (PIGEX):** Technique which looks at product rather than substrate induced gene expression.

***PSI-BLAST* (Position-Specific Iterative BLAST):** an iterative search using the BLAST algorithm. A profile is built after the initial search, which is then used in subsequent searches. The process may be repeated, if desired with new sequences found in each cycle used to refine the profile.

Pyrosequencing: a novel high-throughput nucleotide sequencing method that is based on multiple parallel extensions from target DNA molecules coupled to on-line sensitive reads.

Query**:** The input sequence with which all of the entries in a database are to be compared.

Scaffold: a portion of the genome sequence reconstructed from end-sequenced whole-genome shotgun clones. Scaffolds are composed of a linear ordering (order & orientation) of contigs joined by mate pairs, as well as gaps. Celera Assembler uses complex criteria to build scaffolds, but every sequence gap in the output is spanned by at least two mate pairs.

Sequencing Library**:** Collection of fragments from a genome that are cloned within plasmid vectors or adaptors.

Sequencing Read**:** a contiguous length of nucleotide bases that is generated using a sequencing machine.

***Shotgun sequencing*:** technique in which DNA is broken up randomly into small segments, which are then sequenced using the chain termination or 'pyrosequencing' method to obtain 'reads'. Multiple overlapping reads for the target DNA are obtained and overlapping ends of different reads are assembled by software packages into contiguous sequences denoted 'contigs'.

Similarity: The extent to which nucleotide or protein sequences are related. The extent of similarity between two sequences can be based on percent sequence identity and/or conservation. In BLAST similarity refers to a positive matrix score.

***Singleton*:** It is a read that could not assemble. Singletons can represent contamination, unique sequence with no overlap due to the fluctuation of random coverage, or sequence with so many overlaps it could not be assembled efficiently.

***Stable isotope probing* (SIP):** the use of stable isotopes, such as ^{13}C, as markers to identify organisms that are actively involved in transforming the ^{13}C labeled material (substrate).

***Structural Annotation*:** the notable features by position on a DNA, RNA, or protein sequence. Typically this consists of transcribed sequences, splice junctions, binding sites, functional motifs, active sites, and more.

***Sub surface*:** The geological zone below the surface of the Earth. It is not exposed to the Earth's surface.

***Substrate-induced gene expression screening* (SIGEX):** screening strategy developed for metagenomic libraries in which the induction of gene expression is used as the criterion for rapid identification and isolation of clones.

***Suppressive soil*:** soil that is able to suppress the development of plant diseases by pathogens that are present.

***Suppressive subtractive hybridisation* (SSH):** SSH identifies genetic differences between microorganisms, but the complexity of metagenomes makes this detection difficult. SSH has successfully been used on complex metagenomes and the sensitivity of the process can be increased by using multiple rounds. These techniques have been effectively applied for eukaryotic gene. Their high sensitivity and selectivity should enable small differences in expression of single copy genes to be detected.

Transcriptome: *The* full complement of activated genes, mRNAs, or transcripts in a particular sample at a particular time

BIBLIOGRAPHY

Brady, S. F., C. J. Chao, J. Handelsman, and J. Clardy. 2001. Cloning and heterologous expression of a natural product biosynthetic gene cluster from cDNA. *Organic Letters, 3*:1981–1984.

Cantarel, B.L., Coutinho, P.M., Rancurel, C., Bernard, T., Lombard, V., Henrissat, B.,2009. The carbohydrate-active enzymes database (CAZy): an expert resource for glycogenomics. *Nucleic Acids Research,* 37: 233–238.

Danie R. 2005. The Metagenomics of Soil. *Nature reviews,(3) : 470-478*

Delmont ,T.O., P. Robe, S. Cecillon, M. Clark, F. Constancias, P. Simonet, P.R. Hirsch and T. M. Vogel. 2011. Accessing the Soil Metagenome for Studiesof Microbial Diversity. *Applied and Environmental Microbiology.* 77(4):1315.-1324

Dinsdale, E.A., R. A. Edwards, D. Hall, F. Angly., M. Breitbart., J.M. Brulc, M. Furlan, C.Desnues, M. Haynes, L. Li, L. McDaniel, M.A. Moran, K. E. Nelson, C.Nilsson, R. Olson, J. Paul, B. R. Brito, Y. Ruan, B. K. Swan, R. Stevens, D. L. Valentine, R. V. Thurber, L. Wegley, B. A. White and F.Rohwer. 2008. Functional metagenomic profiling of nine biomes. Nature, 452:629-632.

Đokiæ, L., M. Saviæ, T. Naranèiæ and B. Vasiljeviæ. 2010. Metagenomic Analysis Of Soil Microbial Communities. *Archives of Biological Science*, , 62 (3):559-564.

Fierer,N., M. Breitbart, J. Nulton, P. Salamon, A. Edwards, B. Felts, S. Rayhawk, R. Knight, F. Rohwer and R. B. Jackson. 2007. Metagenomic and Small-Subunit rRNA Analyses Reveal the Genetic Diversity of Soil Bacteria, Archaea, Fungi, and Viruses. *Applied and Environmental Microbiology.* 73(21):7059. DOI:

Gillespie, D. E., S. F. Brady, A. D. Bettermann, N. P. Cianciotto, M. R. Liles, M. R. Rondon, J. Clardy, R. M. Goodman, and J. Handelsman. 2002. Isolation of antibiotics turbomycin A and B from a metagenomic library of soil microbial DNA. *Applied and Environmental Microbiology, 68* :4301–4306.

Handelsman, J. 2004. Metagenomics: Applicayion of genomics to uncultured microorganisms. *Microbiology and Molecular Biology Review, 68* : 669 - 685

Huggett, J. F., Laver, T., Tamisak, S., Nixon, G.,Denise M., O'Sullivan, Elaswarapu, R. and D.J Studholme. 2013. Considerations for the development and application of control materials to improve metagenomic microbial community profiling. *Accred Qual Assur 18*:77–83

J. Xu. 2006. Microbial ecology in the age of genomics and metagenomics: concepts, tools, and recent advances. Molecular Ecology, (15) : , 1713–1731

Kakirde, K.S., L.C. Parsley and M.R. Liles. 2010. Size Does Matter: Application-driven Approaches for Soil Metagenomics. *Soil Biology and Biochemistry.* 42(11): 1911–1923.

Knietsch, K., T. Waschkowitz, S. Bowien, A. Henne and R. Daniel.2003. Construction and Screening of Metagenomic Libraries Derived from Enrichment Cultures: Generation of a Gene Bank for Genes Conferring Alcohol Oxidoreductase Activity on *Escherichia coli. Applied and Environmental Microbiology, 69(3):1408-1416.*

Lindahl, V.,and L. R. Bakken. 1995. Evaluation of methods for extraction of bacteria from soil FEMS Microbiology Ecology ,(16): 135-142

MacNeil, I.A., C.L. Tiong, C. Minor, P.R. August, T.H. Grossman, K.A. Loiacono, B.A Lynch, T. Phillips and S. Narula. 2001. Expression and isolation of antimicrobial small molecules from soil DNA libraries *Journal of Molecular Microbiology and Biotechnology, 3*:301-308

Magnani, M., L. Galluzzi, I.J. and Bruce. 2006. The use of magnetic nanoparticles in the development of new molecular detection systems. *J Nanosci Nanotechnology* ,6: 2302–2311.

Nacke, H., Engelhaupt, M., Brady, S., Fischer, C., Tautzt, J., Daniel, R., 2012. Identification and characterization of novel cellulolytic and hemicellulolytic genesand enzymes derived from German grassland soil metagenomes. *Biotechnology Letters*, 34: 663–675.

Pope, P.B., Denman, S.E., Jones, M., Tringe, S.G., Barry, K., Malfatti, S.A., McHardy, A.C.,Cheng, J.-F., Hugenholtz, P., McSweeney, C.S., Morrison, M., 2010. Adaptation toherbivory by the Tammar wallaby includes bacterial and glycoside hydrolaseprofiles different from other herbivores. *Proceedings of the National Academyof Sciences,* 107:14793–14798.

Rhee, J.K., D.G.Ahn, Y.G.Kim and J.W.Oh. 2005. New thermophilic and thermostable esterase with sequence similarity to the hormone sensitive lipase family, cloned from a metagenomic library. *Applied and Environmental Microbiology, 71*:817-825.

Rondon,M.R., P.R.August,A.D. Bettermann, S.F. Brady, T.H.Grossman, M.R.Liles and K.A.Loiacono 2000. Cloning the soil metagenome: a strategy for accessing the genetic and functional diversity of uncultured microorganisms. *Applied and Environmental Microbiology, 66*:2541-2547

Rondon, M.R., P.R.August, A. D. Bettermann, S.F. Brady,T. H. Grossman,M. R. Liles,K. A. Loiacono, B. A. Lynch, I.A. Macneil, C. Minor,C.L. Tiong, M. GILMAN,M.S. Osburne,J. Clardy, J. Handelsman and R.M. Goodman. 2000.Cloning the Soil Metagenome: a Strategy for Accessing the Genetic and Functional Diversity of Uncultured Microorganisms. Applied and Environmental Microbiology, 66 (6): 2541–2547

Sebastianelli, T. Sen and I.J. Bruce. 2007.Extraction of DNA from soil using nanoparticles by magnetic bioseparation. *Letters in Applied Microbiology*, 46:488-491

Schloss, P.D. and J. Handelsman. 2003. Biotechnological prospects from metagenomics.Current Opinion in Biotechnology, 14:303–310

Thurber, R.V., M. Haynes, M. Breitbart, L. Wegley and F.Rohwer. 2009. Laboratory procedures to generate viral metagenomes. Nature Protocols, 4(4):470-483

Uchiama, T., T.Abe, T.Ikemura and K. Watanabe. 2005. Substrate induced gene expression of environmental metagenome libraries for isolation of catabolic genes. *Nature Biotechnology, 23* : 88-93

Uchiyama, T and K. Miyazaki. 2009.Functional metagenomics for enzyme discovery: challenges to efficient screening. *Current Opinion in Biotechnology*, 20:616–622

Uchiyama, T and K. Miyazaki. 2009.Functional metagenomics for enzyme discovery: challenges to efficient screening. Current Opinion in Biotechnology, 20:616–622

Voget, S., C. Leggewie, A. Uesbeck, C. Raasch, K.E. Jaeger and W.R. Streit. 2003. Prospecting for novel biocatalysts in a soil metagenome. *Applied and Enviromental Microbiology, 69* : 6235-6242.

Yun, J., S.Kang, S.Park, H.Yoon, M.J.Kim, S.Heu and S.Ryu. 2004. Characterization of a novel amylolytic enzyme encoded by a gene from a soil derived metagenomic library. *Applied and Environmental Microbiology, 70*:7229-7235.